Handbook of Tunnel Engineering Design Construction and Risk Assessment

Handbook of
Tunnel Engineering Design
Construction and Risk Assessment

Brian Gayle
Richard Gayle

KNOWLEDGE BAKERS

Cataloging in Publication Data--DK
Courtesy: D.K. Agencies (P) Ltd. <docinfo@dkagencies.com>

Gayle, Brian, author.
Handbook of tunnel engineering design : construction and risk assessment / Brian Gayle, Richard Gayle.
pages cm
Includes bibliographical references and index.
ISBN 978-93-49964-71-6

1. Tunnels--Design and construction. 2. Tunnels--Risk assessment. I. Gayle, Richard, author. II. Title.

LCC TA805.G39 2023 | DDC 624.193 23

ISBN: 978-93-49964-71-6

Published by : **Knowledge Bakers**
Pune, India
Email: knowledgebakers2019@gmail.com

Preface

At its most basic, a tunnel is a tube hollowed through soil or stone. Constructing a tunnel, however, is one of the most complex challenges in the field of civil engineering. Many tunnels are considered technological masterpieces and governments have honoured tunnel engineers as heroes. That's not to say, of course, that some tunnel projects haven't encountered major setbacks. A tunnel is a horizontal passageway located underground. While erosion and other forces of nature can form tunnels, in this article we'll talk about man made tunnels — tunnels created by the process of excavation. There are many different ways to excavate a tunnel, including manual labour, explosives, rapid heating and cooling, tunnelling machinery or a combination of these methods. Some structures may require excavation similar to tunnel excavation, but are not actually tunnels. Shafts, for example, are often hand-dug or dug with boring equipment. But unlike tunnels, shafts are vertical and shorter. Often, shafts are built either as part of a tunnel project to analyse the rock or soil, or in tunnel construction to provide headings, or locations, from which a tunnel can be excavated. There are three broad categories of tunnels: mining, public works and transportation. Let's look briefly at each type.

Mine tunnels are used during ore extraction, enabling labourers or equipment to access mineral and metal deposits deep inside the earth. These tunnels are made using similar techniques as other types of tunnels, but they cost less to build. Mine tunnels are not as safe as tunnels designed for permanent occupation, however. Public works tunnels carry water, sewage or gas lines across great distances. The earliest tunnels were used to transport water to, and sewage away from, heavily populated regions. Roman engineers used an extensive network of tunnels to help carry water from mountain springs to cities and villages. By A.D. 97, nine aqueducts carried approximately 85 million gallons of water a day from mountain springs to the city of Rome. Before there were trains and cars, there were transportation tunnels such as canals — artificial waterways used for travel, shipping or irrigation. Just like railways and roadways today, canals usually ran above ground, but many required tunnels to pass efficiently through an obstacle, such as a mountain. Canal construction inspired some of the world's earliest tunnels. The Underground Canal, located in Lancashire County and Manchester, England, was constructed from the mid- to late-1700s and includes miles of tunnels to house the underground canals. One of America's first tunnels was the Paw Paw Tunnel, built in West Virginia between 1836 and 1850 as part of the Chesapeake and Ohio Canal. Although the canal no longer runs through the Paw Paw, at 3,118 feet long it is still one of the longest canal tunnels in the United States. By the 20th century, trains and cars had replaced canals as the primary form of transportation, leading to the construction of bigger, longer tunnels. The Holland Tunnel, completed in 1927, was one of the first roadway tunnels and is still one of the world's greatest engineering projects. Named for the engineer who oversaw construction, the tunnel ushers nearly 100,000 vehicles daily between New York City and New Jersey. Almost every tunnel is a solution to a specific challenge or problem. In many cases, that challenge is an obstacle that a roadway or railway must bypass. For New York City officials, the solution was clear: Build an automobile tunnel under the river and let commuters drive themselves from New Jersey into the city. The tunnel made an immediate impact. On the opening day alone, 51,694 vehicles made the crossing, with an average trip time of just 8 minutes. Sometimes, tunnels offer a safer solution than other structures. The Seikan Tunnel in Japan was built because ferries crossing the Tsugaru Strait often encountered dangerous waters and weather conditions. After a typhoon sank five ferryboats in 1954, the Japanese government considered a variety of solutions. Ten years later, construction began, and in 1988, the Seikan Tunnel officially opened. How a tunnel is built depends heavily on the material through which it must pass. Tunnelling through soft ground, for instance, requires very different techniques than tunnelling through hard rock or soft rock, such as shale, chalk or sandstone. Tunnelling underwater, the most challenging of all environments, demands a unique approach that would be impossible or impractical to implement above ground.

The Handbook will take profound knowledge and a versatile mind to master the disciplines involved including mechanics, material science, energy technology, heat and mass transfer, hydraulics, acoustics, risk assessment, computer modelling and simulation, 3D-visualization, communication, and management.

Dr Brian Gayle
Dr Richard Gayle

Contents

Chapter 1

Introduction

Tunnel

A tunnel is an underground passageway, completely enclosed except for openings for egress, commonly at each end.

Figure: Underground railway tunnel on the Taipei Metro in Taiwan.

A tunnel may be for foot or vehicular road traffic, for rail traffic, or for a canal. Some tunnels are aqueducts to supply water for consumption or for hydroelectric stations or are sewers. Other uses include routing power or telecommunication cables, some are to permit wildlife such as European badgers to cross highways. Secret tunnels have given entrance to or escape from an area, such as the Cu Chi Tunnels or the smuggling tunnels in the Gaza Strip which connect it to Egypt. Some tunnels are not for transport at all but rather, are fortifications, for example Mittelwerk and Cheyenne Mountain.

In the United Kingdom, a pedestrian tunnel or other underpass beneath a road is called a underpass subway. In the United States that term now means an underground rapid transit system. The central part of a rapid transit network is usually built in tunnels. Rail station platforms may be connected by pedestrian tunnels or by foot bridges.

Figure: Coal tunnel under construction using handset aluminum concrete forms.

A tunnel is relatively long and narrow; in general the length is more (usually much more) than twice the diameter, although similar shorter excavations can be constructed such as cross passages between tunnels.

The definition of what constitutes a tunnel can vary widely from source to source. For example the definition of a road tunnel in the United Kingdom is defined as "a subsurface highway structure enclosed for a length of 150m, or more." In the United States the NFPA definition of a tunnel is "An underground structure with a design length greater than 23 m (75 ft) and a diameter greater than 1800 mm (6 ft)

Geotechnical Investigation

A tunnel project must start with a comprehensive investigation of ground conditions by collecting samples from boreholes and by other geophysical techniques.

An informed choice can then be made of machinery and methods for excavation and ground support, which will reduce the risk of encountering unforeseen ground conditions. In planning the route the horizontal and vertical alignments will make use of the best ground and water conditions.

In some cases conventional desk and site studies yield insufficient information to assess such factors as the blocky nature of rocks, the exact location of fault zones, or the stand-up times of softer ground. This may be a particular concern in large diametre tunnels.

To give more information a pilot tunnel, or drift, may be driven ahead of the main drive.

This smaller diametre tunnel will be easier to support should unexpected conditions be met, and will be incorporated in the final tunnel. Alternatively, horizontal boreholes may sometimes be drilled ahead of the advancing tunnel face.

Construction

Figure: Cut-and-cover constructions of the Paris Métro in France

Tunnels are dug in types of materials varying from soft clay to hard rock. The method of tunnel construction depends on such factors as the ground conditions, the ground water conditions, the length and diametre of the tunnel drive, the depth of the tunnel, the logistics of supporting the tunnel excavation, the final use and shape of the tunnel and appropriate risk management.

There are three basic types of tunnel construction in common use:

- Cut and cover tunnels, constructed in a shallow trench and then covered over.
- Bored tunnels, constructed in situ, without removing the ground above. They are usually of circular or horseshoe cross-section.

- Immersed tube tunnels, sunk into a body of water and sit on, or are buried just under, its bed.

Figure: Cutting shield used for the New Elbe Tunnel.

Cut-and-Cover

Cut-and-cover is a simple method of construction for shallow tunnels where a trench is excavated and roofed over with an overhead support system strong enough to carry the load of what is to be built above the tunnel. Two basic forms of cut-and-cover tunnelling are available:

- *Bottom-up method*: A trench is excavated, with ground support as necessary, and the tunnel is constructed in it. The tunnel may be of in situ concrete, precast concrete, precast arches,or corrugated steel arches; in early days brickwork was used. The trench is then carefully back-filled and the surface is reinstated.
- *Top-down method*: Here side support walls and capping beams are constructed from ground level by such methods as slurry walling, or contiguous bored piling. Then a shallow excavation allows making the tunnel roof of precast beams or in situ concrete. The surface is then reinstated except for access openings. This allows early reinstatement of roadways, services and other surface features. Excavation then takes place under the permanent tunnel roof, and the base slab is constructed.

Shallow tunnels are often of the cut-and-cover type (if under water, of the immersed-tube type), while deep tunnels are excavated, often using a tunnelling shield. For intermediate levels, both methods are possible.

Large cut-and-cover boxes are often used for underground metro stations, such as Canary Wharf tube station in London. This construction form generally has two levels, which allows economical arrangements for ticket hall, station platforms, passenger access and emergency egress, ventilation and smoke control, staff rooms, and equipment rooms.

The interior of Canary Wharf station has been likened to an underground cathedral, owing to the sheer size of the excavation. This contrasts with most traditional stations on London Underground, where bored tunnels were used for stations and passenger access.

Open Building Pit

An open building pit consists of a horizontal and a vertical boundary that keeps groundwater and soil out of the pit. There are several potential alternatives and combination's for (horizontal and vertical) building pit boundaries. The most important difference with Cut-and-cover is that the open building pit is muted after tunnel construction, no roof is placed.

Clay-Kicking

Clay-kicking is a specialised method developed in the United Kingdom, of manually digging tunnels in strong clay-based soil structures. Unlike previous manual methods of using mattocks which relied on the soil structure to be hard, clay-kicking was relatively silent and hence did not harm soft clay based structures.

The clay-kicker lies on a plank at a 45-degree angle away from the working face, and inserts a tool with a cup-like rounded end with his feet. Turning the tool with his hands, he extracts a section of soil, which is then placed on the waste extract.

Regularly used in Victorian civil engineering, the methods found favour in the renewal of the United Kingdom's then ancient sewerage systems, by not having to remove all property or infrastructure to create an effective small tunnel system. During the First World War, the system was successfully deployed by the Royal Engineer tunnelling companies to deploy large military mines beneath enemy German Empire lines. The method was virtually silent not susceptible to listening methods of detection.

Boring Machines

Tunnel boring machines (TBMs) and associated back-up systems are used to highly automate the entire tunneling process, reducing tunneling costs.

Figure 3: A tunnel boring machine that was used at Yucca Mountain nuclear waste repository

Tunnel boring in certain predominantly urban applications, is viewed as quick and cost effective alternative to laying surface rails and roads. Expensive compulsory purchase of buildings and land with potentially lengthy planning inquiries is eliminated.

There are a variety of TBMs that can operate in a variety of conditions, from hard rock to soft water-bearing ground. Some types of TBMs, bentonite slurry and earth-pressure balance machines, have pressurised compartments at the front end, allowing them to be used in difficult conditions below the water table.

This pressurizes the ground ahead of the TBM cutter head to balance the water pressure. The operators work in normal air pressure behind the pressurised compartment, but may occasionally have to enter that compartment to renew or repair the cutters. This requires special precautions, such as local ground treatment or halting the TBM at a position free from water. Despite these difficulties, TBMs are now preferred to the older method of tunneling in compressed air, with an air lock/decompression chamber some way back from the TBM, which required operators to work in high pressure and go through decompression procedures at the end of their shifts, much like divers.

In February 2010, Aker Wirth delivered a TBM to Switzerland, for the expansion of the Linth–Limmern Power Stations in Switzerland. The borehole has a diameter of 8.03 metres (26.3 ft). The TBM used for digging the 57-kilometre (35 mi) Gotthard Base Tunnel, in Switzerland, has a diameter of about 9 metres (30 ft). A larger TBM was built to bore the Green Heart Tunnel (Dutch: Tunnel Groene Hart) as part of the HSL-Zuid in the Netherlands, with a diameter of 14.87 metres (48.8 ft). This in turn was superseded by the Madrid M30 ringroad, Spain, and the Chong Ming tunnels in Shanghai, China. All of these machines were built at least partly by Herrenknecht.

Shafts

A shaft is sometimes necessary for a tunnel project. They are usually circular and go straight down until they reach the level at which the tunnel is going to be built. A shaft normally has concrete walls and is built to be permanent. Once they are built, Tunnel Boring Machines are lowered to the bottom and excavation can start. Shafts are the main entrance in and out of the tunnel until the project is completed. Sometimes if a tunnel is going to be long, multiple shafts at various locations will be bored so that entrance into the tunnel is closer to the unexcavated area.

Other Key Factors

Stand-up time is the amount of time a tunnel will support itself without any added structures. Knowing this time allows the engineers to determine how much can be excavated before support is needed. The longer the stand-up time is the faster the excavating will go. Generally certain configurations of rock and clay will have the greatest stand-up time, and sand and fine soils will have a much lower stand-up time.

Groundwater control is very important in tunnel construction. If there is water leaking into the tunnel stand-up time will be greatly decreased. If there is water leaking into the shaft it will become unstable and will not be safe to work in. To stop this from happening there are a few common methods. One of the most effective is ground freezing. To do this pipes are inserted into the ground surrounding the shaft and are cooled until they freeze. This freezes the ground around each pipe until the whole shaft is surrounded frozen soil, keeping water out. The most common method is to install pipes into the ground and to simply pump the water out. This works for tunnels and shafts.

Tunnel shape is very important in determining stand-up time. The force from gravity is straight down on a tunnel, so if the tunnel is wider than it is high it will have a harder time supporting itself, decreasing its stand-up time. If a tunnel is higher than it is wide the stand up time will increase making the project easier. The hardest shape to support itself is a square or rectangular tunnel. The forces have a harder time being redirected around the tunnel making it extremely hard to support itself. This of course all depends what the material of the ground is.

Tunnel Boring Machine

Figure: Top view of a model of the TBM used on the Gotthard Base Tunnel.

A tunnel boring machine (TBM) also known as a "mole", is a machine used to excavate tunnels with a circular cross section through a variety of soil and rock strata. They can bore through anything from hard rock to sand. Tunnel diametres can range from a metre (done with micro-TBMs) to almost 16 metres to date. Tunnels of less than a metre or so in diametre are typically done using trenchless construction methods or horizontal directional drilling rather than TBMs. Tunnel boring machines are used as an alternative to drilling and blasting (D&B) methods in rock and conventional "hand mining" in soil. TBMs have the advantages of limiting the disturbance to the surrounding ground and producing a smooth tunnel wall. This significantly reduces the cost of lining the tunnel, and makes them suitable to use in heavily urbanised areas. The major disadvantage is the upfront cost. TBMs are expensive to construct, and can be difficult to transport. However, as modern tunnels become longer, the cost of tunnel boring machines versus drill and blast is actually less—

this is because tunnelling with TBMs is much more efficient and results in a shorter project.

Figure: Hydraulic jacks holding a TBM in place.

The largest diametre TBM, at 19.25 m, built by Herrenknecht AG for a recent project in Orlovski Tunnel, St.Petersburg. The machine was built to bore through soft ground including sand and clay. The largest diametre hard rock TBM, at 14.4 m, was manufactured by "The Robbins Company for Canada's Niagara Tunnel Project". The machine was used to bore a hydroelectric tunnel beneath Niagara Falls, the machine has been named "Big Becky" in reference to the Sir Adam Beck hydroelectric dams to which it is tunnelling to provide an additional hydroelectric tunnel.

History

The first successful tunnelling shield was developed by Sir Marc Isambard Brunel to excavate the Thames Tunnel in 1825. However, this was only the invention of the shield concept and did not involve the construction of a complete tunnel boring machine, the digging still having to be accomplished by the then standard excavation methods.

The first boring machine reported to have been built was Henri-Joseph Maus' *Mountain Slicer*. Commissioned by the King of Sardinia in 1845 to dig the Fréjus Rail Tunnel between France and Italy through the Alps, Maus had it built in 1846 in an arms factory near Turin. It consisted of more than 100 percussion drills mounted in the front of a locomotive-sized machine, mechanically power-driven from the entrance of the tunnel. The Revolutions of 1848 affected the funding, and the tunnel was not completed until 10 years later, by using innovative but less expensive methods such as pneumatic drills.

In the United States, the first boring machine to have been built was used in 1853 during the construction of the Hoosac Tunnel. Made of cast iron, it was known as *Wilson's Patented Stone-Cutting Machine*, after inventor Charles Wilson. It drilled 10 feet into the rock before breaking down. The tunnel was eventually completed more than 20 years later, and as with the Fréjus Rail Tunnel, by using less ambitious methods.

In the early 1950s, F.K. Mitry won a dam diversion contract for the Oahe Dam in Pierre, South Dakota, and consulted with James S. Robbins, founder of The Robbins Company, to dig through what was the most difficult shale to excavate at that time, the Pierre Shale. Robbins built a machine that was able to cut 160 feet in 24 hours in the shale, ten times faster than any other method at that time.

The breakthrough that made tunnel boring machines efficient and reliable was the invention of the rotating head mounted with disc cutters. Initially, Robbins' tunnel boring machine used steel picks rotating in a circular motion to dig the excavation front, but he quickly discovered that these picks, no matter how strong they were, had to be changed frequently as they broke or tore off. By replacing the picks with longer lasting disc cutters, this problem was significantly reduced. The design was first utilised successfully at the Humber River Sewer Tunnel in 1956 (Foley, 2009). Since then, all successful hard rock tunnel boring machines have utilised rotating cutting wheels with circular disc cutters.

Description

Modern TBMs typically consist of the rotating cutting wheel, called a cutter head, followed by a main bearing, a thrust system and trailing support mechanisms. The type of machine used depends on the particular geology of the project, the amount of ground water present and other factors.

Hard Rock TBMs

In hard rock, either shielded or open-type TBMs can be used. All types of hard rock TBMs excavate rock using disc cutters mounted in the cutter head. The disc cutters create compressive stress fractures in the rock, causing it to chip away from the rock in front of the machine, called the tunnel face. The excavated rock, known as muck, is transferred through openings in the cutter head to a belt conveyor, where it runs through the machine to a system of conveyors or muck cars for removal from the tunnel.

Open-type TBMs have no shield, leaving the area behind the cutter head open for rock support. To advance, the machine uses a gripper system that pushes against the side walls of the tunnel. Not all machines can be continuously steered while gripper shoes push on the side-walls, As in the case of a Wirth machine which will only steer while ungripped.

The machine will then push forward off the grippers gaining thrust, At the end of a stroke, the rear legs of the machine are lowered, the grippers and propel cylinders are retracted. The retraction of the propel cylinders repositions the gripper assembly for the next boring cycle. The grippers are extended, the rear legs lifted, and boring begins again. The open-type, or Main Beam, TBM does not install concrete segments behind it as other machines do. Instead, the rock is held up using ground support methods such as ring beams, rock bolts, shotcrete, steel straps, Ring steel (Pat 2011) and wire mesh (Stack, 1995).

In fractured rock, shielded hard rock TBMs can be used, which erect concrete segments to support unstable tunnel walls behind the machine. Double Shield TBMs have two modes; in stable ground they can grip against the tunnel walls to advance. In unstable, fractured ground, the thrust is shifted to thrust cylinders that push off against the tunnel segments behind the machine. This keeps the significant thrust forces from impacting fragile tunnel walls. Single Shield TBMs operate in the same way, but are used only in fractured ground, as they can only push off against the concrete segments (Stack, 1995).

Soft Ground TBMs

In soft ground, there are three main types of TBMs: Earth Pressure Balance Machines (EPB), Slurry Shield (SS) and open-face type. Both types of closed machines operate like Single Shield TBMs, using thrust cylinders to advance forward by pushing off against concrete segments. Earth Pressure Balance Machines are used in soft ground with less than 7 bar of pressure. The cutter head does not use disc cutters only, but instead a combination of tungsten carbide cutting bits, carbide disc cutters, and/or hard rock disc cutters. The EPB gets its name because it is capable of holding up soft ground by maintaining a balance between earth and pressure. The TBM operator and automated systems keep the rate of soil removal equal to the rate of machine advance. Thus, a stable environment is maintained. In addition, additives such as bentonite, polymers and foam are injected into the ground to further stabilise it. In soft ground with very high water pressure and large amounts of ground water, Slurry Shield TBMs are needed. These machines offer a completely enclosed working environment. Soils are mixed with bentonite slurry, which must be removed from the tunnel through a system of slurry tubes that exit the tunnel. Large slurry separation plants are needed on the surface for this process, which separate the dirt from the slurry so it can be recycled back into the tunnel.

Open face TBMs in soft ground rely on the fact that the face of the ground being excavated will stand up with no support for a short period of time - this makes them suitable for use in rock types with a strength of up to 10MPa or so, and with low water inflows. Face sizes in excess of 10 metres can be excavated in this manner. The face is excavated using a backactor arm or cutter head to within 150mm of the edge of the shield. The shield is jacked forwards and cutters on the front of the shield cut the remaining ground to the same circular shape. Ground support is provided by use of precast concrete, or occasionally SGI (Spheriodal Graphite Iron), segments that are bolted or supported until a full ring of support has been erected. A final segment, called the key, is wedge-shaped, and expands the ring until it is tight against the circular cut of the ground left behind by cutters on the TBM shield. Many variations of this type of TBM exist.

While the use of TBMs relieves the need for large numbers of workers at high pressures, a caisson system is sometimes formed at the cutting head for slurry shield TBMs. Workers entering this space for inspection, maintenance and repair need to be medically cleared as "fit to dive" and trained in the operation of the locks.

Back-up Systems

Behind all types of tunnel boring machines, inside the finished part of the tunnel, are trailing support decks known as the back-up system. Support mechanisms located on the back-up can include: conveyors or other systems for muck removal, slurry pipelines if applicable, control rooms, electrical systems, dust removal, ventilation and mechanisms for transport of pre-cast segments.

Urban Tunnelling and Near Surface Tunnelling

Urban tunnelling has the special requirement that the ground surface be undisturbed. This means that ground subsidence must be avoided. The normal method of doing this in soft ground is to maintain the soil pressures during and after the tunnel construction. There is some difficulty in doing this, particularly in varied strata (e.g., boring through a region where the upper portion of the tunnel face is wet sand and the lower portion is hard rock).

TBMs with positive face control, such as EPB and SS, are used in such situations. Both types (EPB and SS) are capable of reducing the risk of surface subsidence and voids if operated properly and if the ground conditions are well documented.

When tunnelling in urban environments, other tunnels, existing utility lines and deep foundations need to be addressed in the early planning stages. The project must accommodate measures to mitigate any detrimental effects to other infrastructure.

Shafts

A shaft is sometimes necessary for a tunnel project. They are usually circular and go straight down until they reach the level at which the tunnel is going to be built. A shaft normally has concrete walls and is built to be permanent. Once they are built, Tunnel Boring Machines are lowered to the bottom and excavation can start. Shafts are the main entrance in and out of the tunnel until the project is completed. Sometimes if a tunnel is going to be long, multiple shafts at various locations will be bored so that entrance into the tunnel is closer to the unexcavated area.

Other Key Factors

- Stand-up time is the amount of time a tunnel will support itself without any added structures. Knowing this time allows the engineers to determine how much can be excavated before support is needed. The longer the stand-up time is the faster the excavating will go. Generally certain configurations of rock and clay will have the greatest stand-up time, and sand and fine soils will have a much lower stand-up time.
- Groundwater control is very important in tunnel construction. If there is water leaking into the tunnel stand-up time will be greatly decreased. If there is water leaking into the shaft it will become unstable and will not be safe to work in. To stop this from happening there are a few common methods. One of the most effective is ground freezing. To do this pipes are inserted into the ground surrounding the shaft and are cooled

until they freeze. This freezes the ground around each pipe until the whole shaft is surrounded frozen soil, keeping water out. The most common method is to install pipes into the ground and to simply pump the water out. This works for tunnels and shafts.

- Tunnel shape is very important in determining stand-up time. The force from gravity is straight down on a tunnel, so if the tunnel is wider than it is high it will have a harder time supporting itself decreasing its stand-up time. If a tunnel is higher than it is wide the stand up time will increase making the project easier. The hardest shape to support itself is a square or rectangular tunnel. The forces have a harder time being redirected around the tunnel making it extremely hard to support itself. This of course all depends what the material of the ground is.

New Austrian Tunnelling Method

The New Austrian Tunnelling method (NATM) was developed between 1957 and 1965 in Austria. It was given its name in Salzburg in 1962 to distinguish it from old Austrian tunnelling approach. The main contributors to the development of NATM were Ladislaus von Rabcewicz, Leopold Müller and Franz Pacher. The main idea is to use the geological stress of the surrounding rock mass to stabilise the tunnel itself.

Many have argued that the New Austrian Tunnelling method was not new or Austrian having been previously used else where in Europe and isn't a tunnelling method as much as a philosophy. This aside NATM has no doubt done much to revolutionise tunnelling and bring it into the 21st century

Principles

The NATM integrates the principles of the behaviour of rock masses under load and monitoring the performance of underground construction during construction. The NATM is not a set of specific excavation and support techniques and has often been referred to as a "design as you go" approach to tunnelling providing an optimized support based on observed ground conditions but more correctly it is a "design as you monitor" approach based on observed convergence and divergence in the lining as well as mapping of prevailing rock conditions.

There are seven features on which NATM is based:

- Mobilisation of the strength of rock mass - The method relies on the inherent strength of the surrounding rock mass being conserved as the main component of tunnel support. Primary support is directed to enable the rock to support itself.
- Shotcrete protection - Loosening and excessive rock deformation must be minimised. This is achieved by applying a thin layer of shotcrete immediately after face advance.
- Measurements - Every deformation of the excavation must be measured. NATM requires installation of sophisticated measurement instrumentation. It is embedded in lining, ground, and boreholes.
- Flexible support - The primary lining is thin and reflects recent strata conditions. Active rather than passive support is used and the tunnel is strengthened not by a thicker concrete lining but by a flexible combination of rock bolts, wire mesh and steel ribs.
- Closing of invert - Quickly closing the invert and creating a load-bearing ring is important. It is crucial in soft ground tunnels where no section of the tunnel should be left open even temporarily.
- Contractual arrangements - Since the NATM is based on monitoring measurements, changes in support and construction method are possible. This is possible only if the contractual system enables those changes.
- Rock mass classification determines support measures - There are several main rock classes for tunnels and corresponding support systems for each. These serve as the guidelines for tunnel reinforcement.

Based on the computation of the optimal cross section, just a thin shotcrete protection is necessary. It is applied immediately behind the Tunnel boring machine, to create a natural load-bearing ring and therefore to minimise the rock's deformation. Additionally, geotechnical instruments are installed to measure the later deformation of excavation. Therefore a monitoring of the stress distribution within the rock is possible.

This monitoring makes the method very flexible, even at surprising changes of the geomechanical rock consistency during the tunnelling work, e.g. by crevices or pit water. Such (usual) problems are not solved by thicker shotcrete, but the reinforcement is done by wired concrete which can be combined with steel ribs or lug bolts. The measured rock properties lead to the appropriate tools for tunnel strengthening. Therefore in the last decade NATM was also applied to soft ground excavations and to tunnels in porous sediments. The flexible NATM technique enables immediate adjustments in the construction details, but this requires a flexible contractual system, too.

Philosophy and Controversial Names

NATM was originally developed for use in the Alps where tunnels are commonly excavated at depth and in high in situ stress conditions. The principles of NATM are fundamental to modern day tunnelling, however most city tunnels are built at shallow depth and need not control the release of the in situ stress, seeking instead to minimise settlement. This has led to a confusion in terminology in that tunnelling engineers use "NATM" to mean different things: some define it as a special technique, but others as a sort of philosophy. Recently the scene has been complicated by new terms and alternative names for certain aspects of NATM. This is partly caused by an increased use of the method in the USA, particularly in soft ground shallow tunnels.

Besides the official name *New Austrian Tunnelling Method* other designations are used, e.g. *Sequential Excavation Method* (SEM) or *Sprayed Concrete Lining* (SCL) are often used in shallower tunnels. In Japan sometimes other names were used, e.g. *Centre Dividing Wall NATM*, or *Cross Diaphragm Method* (both abbreviated as *CDM*), and even *Upper Half Vertical Subdivision method* (UHVS).

The Austrian Society of Engineers and Architects defines "NATM" as a method where the surrounding rock or soil formations of a tunnel are integrated into an overall ring-like support structure. Thus the supporting formations will themselves be part of this supporting structure. However, many engineers use "NATM" whenever shotcrete is proposed for initial ground support of an open-face tunnel. Especially with reference to soft ground, the term NATM can be misleading. As noted by Emit Brown, NATM can refer to both a *design philosophy* and a *construction method.*

Key Features

According to E.Brown (Weblink 2), the key features of the *design philosophy* refer to:

- The strength of the ground around a tunnel is deliberately mobilised to the maximum extent possible.
- Mobilisation of ground strength is achieved by allowing controlled deformation of the ground.
- Initial primary support is installed having load-deformation characteristics appropriate to the ground conditions, and installation is timed with respect to ground deformations.
- Instrumentation is installed to monitor deformations in the initial support system, as well as to form the basis of varying the initial support design and the sequence of excavation.

When NATM is seen as a *construction method*, the key features are:

- The tunnel is sequentially excavated and supported, and the excavation sequences can be varied.
- The initial ground support is provided by shotcrete in combination with fibre or welded-wire fabric reinforcement, steel arches (usually lattice girders), and sometimes ground reinforcement (e.g. soil nails, spiling).
- The permanent support is usually (but not always) a cast-in-place concrete lining.

Some experts note that many of these construction methods were used in the US and elsewhere in soft-ground applications, before NATM was described in the literature.

In an article of 2002, Romero states the major difference between the viewpoints of design and of construction: The deformation of the soil (rem.: at soft-ground tunnels) is not easily 'controlled'. Therefore it can be concluded that the excavation and support planned for sequentially excavated, shotcrete-lined tunnels .. utilises NATM construction methods but not necessarily NATM design methods. These details are less essential at tunnels in solid or fair rock.

Pipe Ramming

Pipe ramming (or pipe jacking in British English) is a trenchless method for installation of steel pipes and casings. Distances up to 30m (150 ft) up long and up to 1,500mm (60-inches) in diametre are feasible, although the method can be used for much longer and larger installations. The method is the most useful for shallow installations under railway lines and roads, where other trenchless methods could cause surface settling or heaving. The majority of installations are horizontal, although the method can be used for vertical installations. The method uses pneumatic percussive blows to drive the pipe through the ground. The leading edge of the pipe is almost always open, and is typically closed only when smaller pipes are being installed. Its shape allows a small overcut (to reduce friction between the pipe and soil and improve load conditions on the pipe) and to direct the soil into the pipe interior instead of compacting it outside the pipe. These objectives are usually achieved by attaching a soil-cutting shoe or special bands to the pipe. Further reduction of friction is typically achieved with lubrication, and different types of bentonite and/or polymers can be used (as in horizontal directional boring) for this purpose. Spoil removal from the pipe can be done after the entire pipe is in place (shorter installations). If the pipe containing the spoil becomes too heavy before the installation is complete, the ramming can be interrupted and the pipe cleaned (longer installations). Spoil can be removed by auger, compressed air or water jetting.

Undersea Tunnel

An undersea tunnel is a tunnel which is partly or wholly constructed under a body of water. They are often used where building a bridge or operating a ferry link is impossible, or to provide competition (or relief) for existing bridges or ferry links. There are many reasons for building an undersea tunnel as opposed to the construction of a bridge or establishment of a ferry link.

Advantages

Compared to Bridges: One such advantage would be that a tunnel would still allow shipping to pass. A low bridge would need to be an opening or swing bridge to allow shipping to pass, which can cause traffic congestion. Conversely, a higher bridge that does allow shipping may be unsightly and opposed by the public. Bridges can also be closed due to harsh weather such as high winds. Another possible advantage is space: the downward ramp leading to a tunnel leaves a smaller footprint compared to the upward ramps required by most bridges. Tunnelling will generate soil that has been excavated and this can be used to create new land, as was done with the soil of the Channel Tunnel.

Compared to Ferry Links

As with bridges, albeit with more chance, ferry links will also be closed during adverse weather. Strong winds, or the tidal limits may also affect the workings of a ferry crossing. Travelling through a tunnel is significantly quicker than travelling using a ferry link, shown by the times for travelling through the Channel Tunnel (75–90 minutes for Ferry and 21 minutes on the Eurostar).

Disadvantages

Compared to Bridges: Tunnels require far higher costs of security and construction than bridges. This may mean that over short distances bridges may be preferred rather than tunnels (for example Dartford Crossing). As stated earlier, bridges may not allow shipping to pass, so solutions such as the Oresund Bridge have been constructed.

Compared to Ferry Links

As with bridges, ferry links are far cheaper to construct and operate than tunnels, but are subject to suspension of services due to severe weather and heavy seas.

List of Notable Examples

- Sydney Harbour Tunnel (2.8 km)
- Transbay Tube
- Seikan Tunnel, world's longest undersea railway tunnel (53.9 km), when non-undersea portions of the tunnel are also measured
- Channel Tunnel, world's longest undersea portion railway tunnel (37.9 km)
- Marmaray, connecting Asia and Europe
- Bømlafjord Tunnel, a road tunnel (7.8 km)
- Tokyo Bay Aqua-Line, world's longest undersea portion road tunnel (9.6 km)
- Eiksund Tunnel (7.7 km), world's deepest undersea road tunnel
- Cross Harbour Tunnel, Hong Kong, a busy road tunnel
- New Elbe Tunnel, Hamburg, Germany, 8-lane road tunnel crossing the Elbe river

Box Jacking

Box jacking is similar to pipe jacking, but instead of jacking tubes, a box shaped tunnel is used. Jacked boxes can be a much larger span than a pipe jack with the span of some box jacks in excess of 20m. A cutting head is normally used at the front of the box being jacked and excavation is normally by excavator from within the box.

Drilling and Blasting

Before the advent of tunnel boring machines, drilling and blasting was the only economical way of excavating long tunnels through hard rock, where digging is not possible. Even today, the method is still used in the construction of tunnels, such as in the construction of the Ltschberg Base Tunnel. The decision whether to construct a tunnel using a TBM or using a drill and blast method includes a number of factors such as:

- Tunnel length
- Managing the risks of variations in ground quality
- Required speed of construction
- The required shape of the tunnel

Tunnel length is a key issue that needs to be addressed because large TBMs for a rock tunnel have a high capital cost, but because they are usually quicker than a drill and blast tunnel the price per metre of tunnel is lower. This means that shorter tunnels tend to be less economical to construct with a TBM and are therefore usually constructed by drill and blast. Managing ground conditions can also have a significant effect on the choice with different methods suited to different hazards in the ground.

History

While drilling and blasting saw limited use in pre-industrial times using gunpowder (such as with the Blue Ridge Tunnel in the United States, built in the 1850s), it was not until more powerful (and safer) explosives, such as dynamite (patented 1867), as well as powered drills were developed, that its potential was fully realised.

Drilling and blasting was successfully used to construct tunnels throughout the world, notably the St. Gotthard Tunnel, the Jungfraubahn and even the longest road tunnel in the world, Lærdalstunnelen, are constructed using this method.

Procedure

As the name suggests, drilling and blasting works as follows:

- A number of holes are drilled into the rock, which are then filled with explosives.
- Detonating the explosive causes the rock to collapse.
- Rubble is removed and the new tunnel surface is reinforced.
- Repeating these steps will eventually create a tunnel.

The positions and depths of the holes (and the amount of explosive each hole receives) are determined by a carefully constructed pattern, which, together with the correct timing of the individual explosions, will guarantee that the tunnel will have an approximately circular cross-section.

Rock Support

As the tunnel is incrementally excavated the roof and sides of the tunnel need to be supported to stop the rock falling into the excavation. The philosophy and methods for rock support vary widely but typical rock support systems can include:

- Rock bolts or rock dowels
- Shotcrete
- Ribs or mining arches and lagging
- Cable bolts
- In-situ concrete

Typically a rock support system would include a number of these support methods, each intended to undertake a specific role in the rock support such as the combination of rock bolting and shotcrete.

Hydraulic Splitter

The hydraulic splitter, also known as rock splitter and darda splitter, is a type of portable hydraulic tool that is used in demolition jobs which involve breaking large blocks of concrete and rocks.

There is also a larger excavator mounted rock splitter from Yamamoto Rock Machine, which is suitable for excavation of large volumes of hard rock where blasting is not practical or allowed.

Hydraulic rock splitters consist of two wedges which are inserted in a pre-drilled hole and a hydraulic cylinder is pushing out a centre wedge between the two side wedges forcing them to separate. Hydraulic splitters are now available at hire shops or another alternative could be Expando non explosive demolition agent.

Costs and Cost Overruns of Tunnels

Tunnels are costly and generally more costly than bridges. Large cost overruns are common in tunnel construction.

Choice of Tunnels vs. Bridges

For water crossings, a tunnel is generally more costly to construct than a bridge. Navigational considerations may limit the use of high bridges or drawbridge spans intersecting with shipping channels, necessitating a tunnel. Bridges usually require a larger footprint on each shore than tunnels. There are actually more codes to follow with bridges than with tunnels. In areas with expensive real estate, such as Manhattan and urban Hong Kong, this is a strong factor in tunnels' favour. Boston's Big Dig project replaced elevated roadways with a tunnel system to increase traffic capacity, hide traffic, reclaim land, redecorate, and reunite the city with the waterfront.

The 1934 Queensway Road Tunnel under the River Mersey at Liverpool, was chosen over a massively high bridge for defence reasons. It was feared aircraft could destroy a bridge in times of war. Maintenance costs of a massive bridge to allow the world's largest ships navigate under was considered higher than a tunnel. Similar conclusions were met for the 1971 Kingsway Tunnel under the River Mersey.

Figure: The Queens–Midtown Tunnel in New York City serves as an example of a water-crossing tunnel built instead of a bridge.

Examples of water-crossing tunnels built instead of bridges include the Holland Tunnel and Lincoln Tunnel between New Jersey and Manhattan in New York City, the Queens-Midtown Tunnel between Manhattan and the borough of Queens on Long Island, and the Elizabeth River tunnels between Norfolk and Portsmouth, Virginia, the 1934 River Mersey road Queensway Tunnel and the Western Scheldt Tunnel, Zeeland, Netherlands.

Other reasons for choosing a tunnel instead of a bridge include avoiding difficulties with tides, weather and shipping during construction (as in the 51.5-kilometre or 32.0-mile Channel Tunnel), aesthetic reasons (preserving the above-ground view, landscape, and scenery), and also for weight capacity reasons (it may be more feasible to build a tunnel than a sufficiently strong bridge).

Figure: Illowra Battery utility tunnel, Port Kembla. One of many bunkers south of Sydney.

Some water crossings are a mixture of bridges and tunnels, such as the Denmark to Sweden link and the Chesapeake Bay Bridge-Tunnel in the eastern United States. There are particular hazards with tunnels, especially from vehicle fires when combustion gases can asphyxiate users, as happened at the Gotthard Road Tunnel in Switzerland in 2001. One of the worst railway disasters ever, the Balvano train disaster, was caused by a train stalling in the Armi tunnel in Italy in 1944, killing 426 passengers.

Variant Tunnel Types

Double-Deck Tunnel: Some tunnels are double-deck, for example the two major segments of the San Francisco – Oakland Bay Bridge (completed in 1936) are linked by a double-deck tunnel, the largest diametre bore tunnel in the world.

At construction this was a combination bidirectional rail and truck pathway on the lower deck with automobiles above, now converted to one-way road vehicle traffic on each deck.

The Lion Rock Tunnel, built in the mid-1960s connecting New Kowloon and Sha Tin within the territory of Hong Kong, carries a motorway and an aqueduct. A recent double-decker tunnel with both decks for motor vehicles is the Fuxing Road Tunnel in Shanghai, China. Cars travel on the two-lane upper deck and heavier vehicles on the single-lane lower. Multipurpose tunnel are tunnels that have more than one purpose.

The SMART Tunnel in Malaysia is the first multipurpose tunnel in the world, as it is used both to control traffic and flood in Kuala Lumpur.

Artificial Tunnels

Over-bridges can sometimes be built by covering a road or river or railway with brick or steel arches, and then levelling the surface with earth. In railway parlance, a surface-level track which has been built or covered over is normally called a covered way. Snow sheds are a kind of artificial tunnel built to protect a railway from avalanches of snow. Similarly the Stanwell Park, New South Wales steel tunnel, on the South Coast railway line, protects the line from rockfalls.

Common utility ducts are human-made tunnels created to carry two or more utility lines underground. Through co-location of different utilities in one tunnel, organisations are able to reduce the costs of building and maintaining utilities.

Figure: CPRR Summit Tunnel was in service from 1868 to 1993.

Hazards

Owing to the enclosed space of a tunnel, fires can have very serious effects on users.

The main dangers are gas and smoke production, with low concentrations of carbon monoxide being highly toxic. Fires killed 11 people in the Gotthard tunnel fire of 2001 for example, all of the victims succumbing to smoke and gas inhalation. Over 400 passengers died in the Balvano train disaster in Italy in 1944, when the locomotive halted in a long tunnel. Carbon monoxide poisoning was the main cause of the horrifying death rate.

Examples of Tunnels

In History

- The World's oldest underwater tunnel is rumoured to be the *Terelek kaya tüneli* under Kýzýl River, a little south of the towns of Boyabat and Duragan in Turkey. Estimated to have been built more than 2000 years ago (possibly 5000), it is assumed to have had a defence purpose.
- The qanat or kareez of Persia is a water management system used to provide a reliable supply of water to human settlements or for irrigation in hot, arid and semi-arid climates. The oldest and largest known qanat is in the Iranian city of Gonabad, which after 2700 years, still provides drinking and agricultural water to nearly 40,000 people. Its main well depth is more than 360 m (1,180 ft), and its length is 45 km (28 mi).
- The Eupalinian aqueduct on the island of Samos (North Aegean, Greece). Built in 520 BC by the ancient Greek engineer Eupalinos of Megara. Eupalinos organised the work so that the tunnel was begun from both sides of Mount Kastro. The two teams advanced simultaneously and met in the middle with excellent accuracy, something that was extremely difficult in that time. The aqueduct was of utmost defensive importance, since it ran underground, and it was not easily found by an enemy who could otherwise cut off the water supply to Pythagoreion, the ancient capital of Samos. The tunnel's existence was recorded by Herodotus (as was the mole and harbour, and the third wonder of the island, the great temple to Hera, thought by many to be the largest in the Greek world).

The precise location of the tunnel was only re-established in the 19th century by German archaeologists. The tunnel proper is 1,030 m long (3,380 ft) and visitors can still enter it Eupalinos tunnel.

- The Via Flaminia, an important Roman road, penetrated the Furlo pass in the Apennines through a tunnel which emperor Vespasian had ordered built in 76-77. A modern road, the SS 3 Flaminia, still uses this tunnel, which had a precursor dating back to the 3rd century BC; remnants of this earlier tunnel (one of the first road tunnels) are also still visible.
- Sapperton Canal Tunnel on the Thames and Severn Canal in England, dug through hills, which opened in 1789, was 3.5 km (2.2 mi) long and allowed boat transport of coal and other goods. Above it runs the Sapperton Long Tunnel which carries the "Golden Valley" railway line between Swindon and Gloucester.
- The 1796 Stoddart Tunnel in Chapel-en-le-Frith in Derbyshire is reputed to be the oldest rail tunnel in the world. Rail wagons were horse-drawn.
- The tunnel was created for the first true steam locomotive, from Penydarren to Abercynon. The Penydarren locomotive was built by Richard Trevithick. The locomotive made the historic journey from Penydarren to Abercynon in 1804. Part of this tunnel can still be seen at Pentrebach, Merthyr Tydfil, Wales. This is arguably the oldest railway tunnel in the world, for self-propelled steam engines on rails.
- The Montgomery Bell Tunnel in Tennessee, a 88 m (289 ft), high water diversion tunnel, 4.50-×-2.45 m high (15-×-8.0 ft), to power a water wheel, was built by slave labour in 1819, being the first full scale tunnel in North America.
- Crown Street Station, Liverpool, 1829. Built by George Stephenson, a single track tunnel 291 yd long (266 m) was bored from Edge Hill to Crown Street to serve the world's first passenger railway station. The station was abandoned in 1836 being too far from Liverpool city centre, with the area converted for freight use. Closed down in 1972, the tunnel is disused. However it is the oldest rail tunnel running under streets in the world.

- The 1.26 mile (2.03 km) 1829 Wapping Tunnel in Liverpool, England, was the first rail tunnel bored under a metropolis. Currently disused since 1972. Having two tracks, the tunnel runs from Edge Hill in the east of the city to the south end Liverpool docks being used only for freight. The tunnel is still in excellent condition and is being considered for reuse by Merseyrail rapid transit rail system, with maybe an underground station cut into the tunnel. The river portal is opposite the new Liverpool Arena being ideal for a serving station. If reused it will be the oldest used underground rail tunnel in the world and oldest part of any underground metro system.
- 1836, Lime St Station tunnel, Liverpool. A two track rail tunnel, 1.13 miles (1,811 m) long was bored under a metropolis from Edge Hill in the east of the city to Lime Street. In the 1880s the tunnel was converted to a deep cutting four tracks wide. The only occurrence of a tunnel being removed. A very short section of the original tunnel still exists at Edge Hill station making this the oldest rail tunnel in the world still in use, and the oldest in use under a street, albeit only one street and one building.
- Box Tunnel in England, which opened in 1841, was the longest railway tunnel in the world at the time of construction. It was dug and has a length of 2.9 km (1.8 mi).
- The 0.75 mile long 1842 Prince of Wales Tunnel, in Shildon near Darlington, England, is the oldest sizable tunnel in the world still in use under a settlement.
- The Thames Tunnel, built by Marc Isambard Brunel and his son Isambard Kingdom Brunel and opened in 1843, was the first underwater tunnel and the first to use a tunnelling shield. Originally used as a foot-tunnel, it was a part of the East London Line of the London Underground until 2007, being the oldest section of the system. From 2010 the tunnel becomes a part of the London Overground system.
- The 2.07 miles (3.34 km) Victoria Tunnel in Liverpool, opened in 1848, was bored under a metropolis. Initially used only for rail freight and later freight and passengers serving the Liverpool ship liner terminal, the tunnel runs from Edge Hill

in the east of the city to the north end Liverpool docks. Used until 1972 it is still in excellent condition, being considered for reuse by the Merseyrail rapid transit rail system. Stations being cut into the tunnel are being considered. Also, reuse by a monorail system from the proposed Liverpool Waters redevelopment of Liverpool's Central Docks has been proposed.

- The oldest underground sections of the London Underground were built using the cut-and-cover method in the 1860s. The Metropolitan, Hammersmith & City, Circle and District lines were the first to prove the success of a metro or subway system. Dating from 1863, Baker Street station is the oldest underground station in the world.
- On June 18, 1868, the Central Pacific Railroad's 1,659-foot (506 m) Summit Tunnel (Tunnel #6) at Donner Pass in the California Sierra Nevada mountains was opened permitting the establishment of the commercial mass transportation of passengers and freight over the Sierras for the first time. It remained in daily use until 1993 when the Southern Pacific Railroad closed it in favour of sending all rail traffic through the 10,322-foot (3,146 m) long Tunnel #41 (aka "The Big Hole") built a mile to the south in 1925.
- The 1882 Col de Tende Road Tunnel, at 3182 metres long, was one of the first long road tunnels under a pass, running between France and Italy.
- The Mersey Railway tunnel opened in 1886 running from Liverpool to Birkenhead under the River Mersey. The Mersey Railway was the world's first deep-level underground railway. By 1892 the extensions on land from Birkenhead Park station to Liverpool Central Low level station gave a tunnel 3.12 miles (5029 m) in length. The under river section is 0.75 miles in length, being the longest underwater tunnel in world in January 1886.
- The rail Severn Tunnel was opened in late 1886, at 4 miles 624 yd (7,008 m) long, although only 2¼ miles (3.62 km) of the tunnel is actually under the river. The tunnel replaced the Mersey Railway tunnel's longest under water record, which it held for less than a year.
- James Greathead, in constructing the City & South London Railway tunnel beneath the Thames, opened in 1890, brought

together three key elements of tunnel construction under water: 1) shield method of excavation; 2) permanent cast iron tunnel lining; 3) construction in a compressed air environment to inhibit water flowing through soft ground material into the tunnel heading.

- St. Clair Tunnel, also opened later in 1890, linked the elements of the Greathead tunnels on a larger scale.
- The 1927 Holland Tunnel was the first underwater tunnel designed for automobiles. This fact required a novel ventilation system.

Longest

- The Delaware Aqueduct in New York USA is the longest tunnel, of any type, in the world at 137 km (85 mi). It is drilled through solid rock.
- The Gotthard Base Tunnel will be the longest rail tunnel in the world at 57 km (35 mi). It will be totally completed in 2017.
- The Seikan Tunnel in Japan is the longest undersea rail tunnel in the world at 53.9 km (33.5 mi), of which 23.3 km (14.5 mi) is under the sea.
- The Channel Tunnel between France and the United Kingdom under the English Channel is the second-longest, with a total length of 50 km (31 mi), of which 39 km (24 mi) is under the sea.
- The Lötschberg Base Tunnel opened in June 2007 in Switzerland was the longest land rail tunnel, with a total of 34.5 km (21.4 mi).
- The Lærdal Tunnel in Norway from Lærdal to Aurland is the world's longest road tunnel, intended for cars and similar vehicles, at 24.5 km (15.2 mi).
- The Zhongnanshan Tunnel in People's Republic of China opened in January 2007 is the world's second longest highway tunnel and the longest road tunnel in Asia, at 18 km (11 mi).
- The longest canal tunnel is the Rove Tunnel in France, over 7.12 km (4.42 mi) long.

Notable

- The Central Artery Tunnel in Boston carries approximately 200,000 vehicles/day.

- The Fredhälls Tunnel in Stockholm, Sweden, and the New Elbe Tunnel in Hamburg, Germany, both with around 150,000 vehicles a day, two of the most trafficked tunnels in the world.
- Gerrards Cross tunnel in Britain is notable in that it is being built over a railway cutting that was dug in the early part of the 20th Century. Thus, arguably, making it the tunnel longest in construction by the cut and cover method. When complete a branch of the Tesco supermarket chain will occupy the space above the railway tunnel.
- Williamson's tunnels in Liverpool, built by a wealthy eccentric are probably the largest underground folly in the world.
- New York City Water Tunnel No. 3, started in 1970, has an expected completion date of 2020.
- The Chicago Deep Tunnel Project is a network of 175 km (109 mi) of tunnels designed to reduce flooding in the Chicago area. Started in the mid 1970s, the project is due to be completed in 2019.
- Moffat Tunnel in Colorado straddles the Continental Divide. The tunnel is 6.2 mi (10.0 km) long and at 9,239 ft (2,816 m) above sea level is the highest railroad tunnel in the United States.
- The Fenghuoshan tunnel on Qinghai-Tibet railway is the world's highest railway tunnel, about 4,905 m (16,093 ft) above sea level.
- The La Linea Tunnel in Colombia, will be (2013) the longest, 8.58 km (5.33 mi), mountain tunnel in South America. It crosses beneath a mountain at 2,500 m (8,202.1 ft) above sea level with six lanes and it has a parallel emergency tunnel. The tunnel is subject to serious groundwater pressure. The tunnel, which is currently under construction, will link Bogotá and its urban area with the coffee-growing region and with the main port on the Colombian Pacific coast.
- The Honningsvåg Tunnel (4.443 km (2.76 mi) long) on European route E69 in Norway is the world's northernmost road tunnel, except for mines (which exist on Svalbard).
- The Eiksund Tunnel on national road Rv 653 in Norway is the world's deepest subsea road tunnel (7,776 m long, with deepest point at -287 metres below the sea level, opened in feb. 2008)

Other Uses

Excavation techniques, as well as the construction of underground bunkers and other habitable areas, are often associated with military use during armed conflict, or civilian responses to threat of attack. The use of tunnels for mining is called drift mining. One of the strangest uses of a tunnel was for the storage of chemical weapons.

Natural Tunnels

- Lava tubes are partially empty, cave-like conduits underground, formed during volcanic eruptions by flowing and cooling lava.
- Natural Tunnel State Park (Virginia, USA) features an 850-foot (259 m) natural tunnel, really a limestone cave, that has been used as a railroad tunnel since 1890.
- Punarjani Guha Kerala, India. Hindus believe that crawling through the tunnel (which they believe was created by a Hindu god) from one end to the other will wash away all of one's sins and thus attain rebirth, although only men are permitted to crawl through the cave.
- Small "snow tunnels" are created by voles, chipmunks and other rodents for protection and access to food sources. For more information regarding tunnels built by animals.

Temporary Way

During construction of a tunnel it is often convenient to install a temporary railway particularly to remove spoil. This temporary railway is often narrow gauge so that it can be double track, which facilitates the operation of empty and loaded trains at the same time. The temporary way is replaced by the permanent way at completion, thus explaining the term Perway.

Enlargement

The vehicles using a tunnel can outgrow it, requiring replacement or enlargement. The original single line Gib Tunnel near Mittagong was replaced with a double line tunnel, with the original tunnel used for growing mushrooms. The Rhyndaston Tunnel was enlarged using a borrowed Tunnel Boring Machine so as to be able to take ISO containers.

The 1836 Lime Street two track 1 mile tunnel from Edge Hill to Lime Street in Liverpool was totally removed, apart from a short 50 metre section at Edge Hill. Four tracks were required. The tunnel was converted into a very deep 4 track open cutting. However, short larger 4 track tunnels were left in some parts of the run. Train services were not interrupted as the work progressed. Photos of the work in progress: There are other occurrences of tunnels being replaced by open cuts, for example, the Auburn Tunnel.

Floor Lowering

Tunnels can also be enlarged by lowering the floor.

Significant Historic and Modern Tunnels

The origin of tunnel building is disputed. The Egyptians built tunnels as entrances to tombs. The Babylonians built (c.2180 B.C.) a tunnel under the Euphrates using what is now called the "cut-and-cover" method; the river was diverted, a wide trench was dug across its bed, and a brick tube was constructed in it and covered up. The ancient Greeks and Romans built tunnels for carrying water and for mining purposes; some of the Roman tunnels are still in use. One of the first notable tunnels in Great Britain was part of the Grand Trunk Canal. It was nearly 2 mi (3.2 km) long and was completed in 1777. The Mont Cénis Tunnel, a railroad tunnel in the French Alps that opened in 1871 and is now 8.5 mi (13.7 km) long, was probably the first tunnel built using compressed-air drills.

The first tunnel of importance in the United States was the tunnel through the Hoosac Range in Massachusetts. There are hundreds of miles of tunnels in New York City and its vicinity, e.g., for subways, roads, water systems, and railroads. The Delaware Aqueduct, which provides part of New York City's water supply, is at 105 mi (168 km) the longest continuous tunnel in the world. Road tunnels include the Holland Tunnel and the Lincoln Tunnel, which connect New York City's Manhattan Island with New Jersey, and the Brooklyn-Battery Tunnel, which connects Manhattan Island with Brooklyn and is longest vehicular tunnel (1.7 mi/2.7 km) in the United States. The Chesapeake Bay Bridge-Tunnel in Virginia, opened in 1964, has a length of 17.6 mi (28.2 km) and includes two tunnel segments over a mile long.

The Simplon Tunnel through the Alps, for many years the longest railway tunnel in the world, consists of two parallel single-line tunnels (both 12.3 mi/19.8 km) with connecting tunnels at short intervals. The

Channel Tunnel (Chunnel) under the English Channel is much longer, at 31 mi (50 km), and the Seikan Tunnel in Japan, the world's deepest underwater tunnel, is also the longest railroad tunnel at 33.5 mi (53.6 km). The world's longest vehicular tunnel, the Lærdal Tunnel (15.2 mi/24.5 km long), connects Lærdal and Aurland, Norway, and is an important overland link between Oslo and Bergen. The St. Gotthard Tunnel (10.2 mi/16.4 km long), in the Swiss Alps, was formerly the longest vehicular tunnel.

Cut-and-Cover Tunnel Method

Cut-and-cover tunnelling is a simple tunnelling construction method used to build shallow tunnels such as those commonly used by subways, railways, and metro systems.Type/Process

Conventional Method: In the conventional method, excavating a trench in the ground and then backfilling and restoring the original roadway or ground is the process used to construct a tunnel. A support system of some sort is also necessary to carry the load of the material used to cover over the tunnel such as shotcrete.

Bottom-up Method

In the cut-and-cover bottom-up or caisson wall method, a drilling rig is used to install caisson walls down to the existing bedrock. Once the caisson walls are in place, soil between the walls is excavated to a depth below the tunnel floor. The tunnel floor, a slab, is poured, followed by the sidewalls of the tunnel from the bottom-up. After the walls of the tunnel are completed, the roof is constructed and the roadway or ground on top of the tunnel restored. Materials used to provide the structure and support in the construction of the tunnel may include concrete, pre-cast concrete, pre-cast arches, or corrugated steel arches.

Top-down Method

In the cut-and-cover top-down or diaphragm wall method, the opposite process takes place in constructing the tunnel. A trencher or trench cutter is typically used to dig a trench out of the ground first before concrete walls are built. This processes consists of using a slurry mixture to build a slurry wall. The slurry wall provides temporary support to the sides of the trench before concrete is poured for a permanent wall structure. Once the concrete walls of the tunnel are completed, the roof of the tunnel is constructed and the surface roadway restored. Excavation of the tunnel is then carried out through openings in the tunnel roof top-down to the tunnel floor. The tunnel floor slab is the last part of construction to be completed.

Cast-in-Place Method

Another type of cut-and-cover tunnelling is called cast-in-place. In this method, a trench is excavated with forms being built directly inside the trench. Concrete is then poured or cast into the concrete. After the concrete cures the forms are removed. The trench is then backfilled and the roadway reinstated. A shoring system is supports the sides of the excavation to prevent the shifting of soil.

Chapter 2

Technical Manual for Design and Construction of Road Tunnels

Planning

Road tunnels are feasible alternatives to cross a water body or traverse through physical barriers such as mountains, existing roadways, railroads, or facilities; or to satisfy environmental or ecological requirements. In addition, road tunnels are viable means to minimise potential environmental impact such as traffic congestion, pedestrian movement, air quality, noise pollution, or visual intrusion; to protect areas of special cultural or historical value such as conservation of districts, buildings or private properties; or for other sustainability reasons such as to avoid the impact on natural habit or reduce disturbance to surface land. The portal for the Glenwood Canyon Hanging Lake and Reverse Curve Tunnels – Twin 4,000 feet (1,219 metre) long tunnels carrying a critical section of I-70 unobtrusively through Colorado's scenic Glenwood Canyon.

Planning for a road tunnel requires multi-disciplinary involvement and assessments, and should generally adopt the same standards as for surface roads and bridge options, with some exceptions as will be discussed later. Certain considerations, such as lighting, ventilation, life safety, operation and maintenance, etc should be addressed specifically for tunnels. In addition to the capital construction cost, a life cycle cost analysis should be performed taking into account the life expectancy of a tunnel. It should be noted that the life expectancies of tunnels are significantly longer than those of other facilities such as bridges or roads.

This chapter provides a general overview of the planning process of a road tunnel project including alternative route study, tunnel type and tunnelling method study, operation and financial planning, and risk analysis and management.

Tunnel Shape and Internal Elements

There are three main shapes of highway tunnels – circular, rectangular, and horseshoe or curvilinear. The shape of the tunnel is largely dependent on the method used to construct the tunnel and on the ground conditions. For example, rectangular tunnels are often constructed by either the cut and cover method, by the immersed method or by jacked box tunnelling). Circular tunnels are generally constructed by using either tunnel boring machine (TBM) or by drill and blast in rock. Horseshoe configuration tunnels are generally constructed using drill and blast in rock or by following the Sequential Excavation Method (SEM), also as known as New Austrian Tunnelling Method (NATM).

Road tunnels are often lined with concrete and internal finish surfaces. Some rock tunnels are unlined except at the portals and in certain areas where the rock is less competent. In this case, rock reinforcement is often needed. Rock reinforcement for initial support includes the use of rock bolts with internal metal straps and mine ties, un-tensioned steel dowels, or tensioned steel bolts. To prevent small fragments of rock from spalling, wire mesh, shotcrete, or a thin concrete lining may be used. Shotcrete, or sprayed concrete, is often used as initial lining prior to installation of a final lining, or as a local solution to instabilities in a rock tunnel. Shotcrete can also be used as a final lining. It is typically placed in layers with welded wire fabric and/or with steel fibres as reinforcement. The inside surface can be finished smooth and often without the fibers. Precast segmental lining is primarily used in conjunction with a TBM in soft ground and sometimes in rock. The segments are usually erected within the tail shield of the TBM. Segmental linings have been made of cast iron, steel and concrete. Presently however, all segmental linings are made of concrete. They are usually gasketed and bolted to prevent water penetration. Precast segmental linings are sometimes used as a temporary lining within which a cast in place final lining is placed, or as the final lining.

Road tunnels are often finished with interior finishes for safety and maintenance requirements. The walls and the ceilings often receive a finish surface while the roadway is often paved with asphalt pavement. The interior finishes, which usually are mounted or adhered to the final lining, consist of ceramic tiles, epoxy coated metal panels, porcelain enameled metal panels, or various coatings. They provide enhanced tunnel lighting and visibility, provide fire protection for the lining, attenuate noise, and provide a surface easy to clean. Design details for final interior finishes are not within the scope of this Manual. The tunnels are usually equipped with various systems such as ventilation, lighting, communication, fire-life safety, traffic operation and control including messaging, and operation and control of the various systems in the tunnel.

Classes of Roads and Vehicle Sizes

A tunnel can be designed to accommodate any class of roads and any size of vehicles. The classes of highways are discussed in A Policy on Geometric Design of Highways and Streets AASHTO (2004). Alignments, dimensions, and vehicle sizes are often determined by the responsible authority based on the classifications of the road (i.e. interstate, state, county or local roads). However, most regulations have been formulated on the basis of open roads. Ramifications of applying these regulations to road tunnels should be considered. For example, the use of full width shoulders in the tunnel might result in high cost. Modifications to these regulations through engineering solutions and economic evaluation should be considered in order to meet the intention of the requirements. The size and type of vehicles to be considered depend upon the class of road. Generally, the tunnel geometrical configuration should accommodate all potential vehicles that use the roads leading to the tunnel including over-height vehicles such as military vehicles if needed. However, the tunnel height should not exceed the height under bridges and overpasses of the road that leads to the tunnel.

On the other hand, certain roads such as Parkways permit only passenger vehicles. In such cases, the geometrical configuration of a tunnel should accommodate the lower vehicle height keeping in mind that emergency vehicles such as fire trucks should be able to pass through the tunnel, unless special low height emergency response vehicles are provided. It is necessary to consider the cost because designing a tunnel facility to accommodate only a very few extraordinary oversize vehicles may not be economical if feasible alternative routes are available. Road tunnel A86 in Paris, for instance, is designed to accommodate two levels of passenger vehicles only and special low height emergency vehicles are provided.

The travelled lane width and height in a tunnel should match that of the approach roads. Often, allowance for repaving is provided in determining the headroom inside the tunnel. Except for maintenance or unusual conditions, two-way traffic in a single tube should be discouraged for safety reasons except like the A-86 Road Tunnel that has separate decks. In addition, pedestrian and cyclist use of the tunnel should be discouraged unless a special duct (or passage) is designed specifically for such use. An example of such use is the Mount Baker Ridge tunnel in Seattle, Washington.

Traffic Capacity

Road tunnels should have at least the same traffic capacity as that of surface roads. Studies suggest that in tunnels where traffic is controlled, throughput is more than that in uncontrolled surface road suggesting that a reduction in the number of lanes inside the tunnel may be warranted. However, traffic will slow down if the lane width is less than standards (too narrow) and will shy away from tunnel walls if insufficient lateral clearance is provided inside the tunnel. Also, very low ceilings give an impression of speed and tend to slow traffic. Therefore, it is important to provide adequate lane width and height comparable to those of the approach road. It is recommended that traffic lanes for new tunnels should meet the required road geometrical requirements (e.g., 12 ft). It is also recommended to have a reasonable edge distance between the lane and the tunnel walls or barriers.

Road tunnels, especially those in urban areas, often have cargo restrictions. These may include hazardous materials, flammable gases and liquids, and over-height or wide vehicles. Provisions should be made in the approaches to the tunnels for detection and removal of such vehicles.

Alternative Analyses

Route Studies: A road tunnel is an alternative vehicular transportation system to a surface road, a bridge or a viaduct. Road tunnels are considered to shorten the travel time and distance or to add extra travel capacity through barriers such as mountains or open waters. They are also considered to avoid surface congestion, improve air quality, reduce noise, or minimise surface disturbance. Often, a tunnel is proposed as a sustainable alternative to a bridge or a surface road. In a tunnel route study, the following issues should be considered:

- Subsurface, geological, and geo-hydraulic conditions
- Constructability
- Long-term environmental impact

- Seismicity
- Land use restrictions
- Potential air right developments
- Life expectancy
- Economical benefits and life cycle cost
- Operation and maintenance
- Security
- Sustainability

Often sustainability is not considered; however, the opportunities that tunnels provide for environmental improvements and real estate developments over them are hard to ignore and should be reflected in term of financial credits. In certain urban areas where property values are high, air rights developments account for a significant income to public agencies which can be used to partially offset the construction cost of tunnels.

It is important when comparing alternatives, such as a tunnel versus a bridge or a bypass, that the comparative evaluation includes the same purpose and needs and the overall goals of the project, but not necessarily every single criterion. For example, a bridge alignment may not necessarily be the best alignment for a tunnel. Similarly, the life cycle cost of a bridge has a different basis than that of a tunnel.

Financial Studies

The financial viability of a tunnel depends on its life cycle cost analysis. Traditionally, tunnels are designed for a life of 100 to 125 years. However, existing old tunnels (over 100 years old) still operate successfully throughout the world. Recent trends have been to design tunnels for 150 years life. To facilitate comparison with a surface facility or a bridge, all costs should be expressed in terms of life-cycle costs. In evaluating the life cycle cost of a tunnel, costs should include construction, operation and maintenance, and financing (if any) using Net Present Value. In addition, a cost-benefit analysis should be performed with considerations given to intangibles such as environmental benefits, aesthetics, noise and vibration, air quality, right of way, real estate, potential air right developments, etc.

The financial evaluation should also take into account construction and operation risks. These risks are often expressed as financial contingencies or provisional cost items. The level of contingencies would be decreased as the project design level advances. The risks are then better quantified and provisions to reduce or manage them are identified.

Types of Road Tunnels

Selection of the type of tunnel is an iterative process taking into account many factors, including depth of tunnel, number of traffic lanes, type of ground traversed, and available construction methodologies. For example, a two-lane tunnel can fit easily into a circular tunnel that can be constructed by a tunnel boring machine (TBM). However, for four lanes, the mined tunnel would require a larger tunnel, two bores or another method of construction such as cut and cover or SEM methods. The maximum size of a circular TBM existing today is about 51 ft (15.43 m) for the construction of Chongming Tunnel, a 5.6 mile (9-kilometre) long tunnel under China's Yangtze River, in Shanghai.

Figure: Chongming Tunnel under the Yangtze River

When larger and deeper tunnels are needed, either different type of construction methods, or multiple tunnels are usually used. For example, if the ground is suitable, SEM in which the tunnel cross section can be made to accommodate multiple lanes can be used. For tunnels below open water, immersed tunnels can be used.

Shallow tunnels would most likely be constructed using cut-and-cover techniques. In special circumstances where existing surface traffic cannot be disrupted, jacked precast tunnels are sometimes used. In addition to the variety of tunnelling methods discussed in this manual, non-conventional techniques have been used to construct very large cross section, such as the Mt. Baker Ridge Tunnel, on I-90 in Seattle, Washington. For that project, multiple overlapping drifts were constructed and filled with concrete to form a circular envelop that provides the overall support system of the ground. Then the space within this envelop was excavated and the tunnel structure was constructed within it.

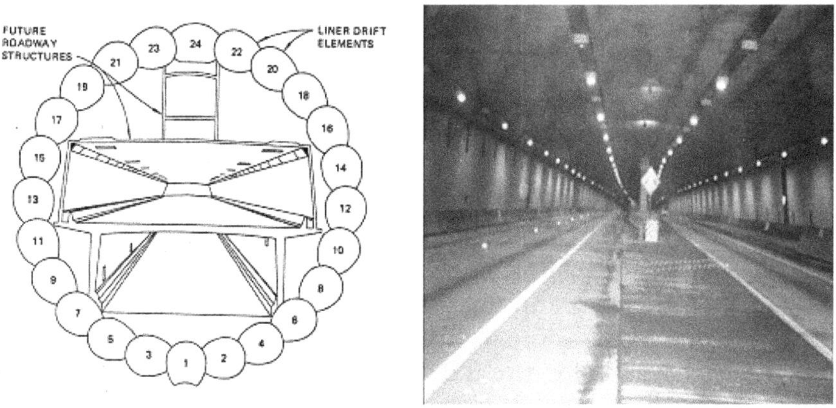

Figure: Stacked Drift and Final Mt Baker Tunnel, I-90, Seattle, WA

There are times when tunnelling is required in a problematic ground such as mixed face (rock and soft ground), squeezing rock or other difficult ground conditions requiring specialised techniques.

Geotechnical Investigations

Geotechnical investigations are critical for proper planning of a tunnel. Selection of the alignment, cross section, and construction methods is influenced by the geological and geotechnical conditions, as well as the site constraints. Good knowledge of the expected geological conditions is essential. The type of the ground encountered along the alignment would affect the selection of the tunnel type and its method of construction. For example, in TBM tunnel construction mixed ground conditions, or buried objects add complications to the TBM performance and may result in the inability of the TBM to excavate the tunnel, potential breakdown of the TBM, or potential ground failure and settlements at the surface. The selection of the tunnel profile must therefore take into account potential ground movements and avoid locations where such movements or settlements could cause surface problems to existing utilities or surface facilities and mitigation measures should be provided.

Another example of the effect of the impact of geological features on the tunnel alignment is the presence of active or inactive faults. During the planning phase, it is recommended to avoid crossing a fault zone and preferred to avoid being in a close proximity of an active fault. However, if avoidance of a fault cannot be achieved, then proper measures for crossing it should be implemented. Special measures may also be required when tunnelling in a ground that may contain methane or other hazardous gasses or fluids.

Geotechnical issues such as the soil or rock properties, the ground water regime, the ground cover over the tunnel, the presence of contaminants along the alignment, presence of underground utilities and obstructions such as boulders or buried objects, and the presence of sensitive surface facilities should be taken into consideration when evaluating tunnel alignment. Tunnel alignment is sometimes changed based on the results of the geotechnical to minimise construction cost or to reduce risks. The tunnel profile can also be adjusted to improve constructability or accommodate construction technologies as long as the road geometrical requirements are not compromised. For example, for TBM tunnels the profile would be selected to ensure that sufficient cover is maintained for the TBM to operate satisfactorily over the proposed length of bore. However, this should not compromise the maximum grade required for the road.

If the route selection is limited, then measures to deal with the poor ground in terms of construction method or ground improvement prior to excavation should be considered. It is recommended that the geotechnical investigation start as early as possible during the initial planning phase of the project. The investigation should address not just the soil and rock properties, but also their anticipated behaviours during excavation. For example in sequential excavation or NATM, ground standup time is critical for its success. If the ground does not have sufficient standup time, pre-support or ground improvement such as grouting should be provided. For soft ground TBM tunnelling, the presence of boulders for example would affect the selection of TBM type and its excavation tools. Similarly, the selection of a rock TBM would require knowledge of the rock unconfined compressive strength, its abrasivity and its jointing characteristics. The investigation should also address groundwater. For example, in soft ground SEM tunnelling, the stability of the excavated face is greatly dependent on control of the groundwater. Dewatering, pre-draining, grouting, or freezing are often used to stabilise the excavation. Ground behaviour during tunnelling will affect potential settlements on the surface. Measures to minimise settlements by using suitable tunnelling methods or by preconditioning the ground to improve its characteristics would be required. Presence of faults or potentially liquefiable materials would be of concern during the planning process. Relocating the tunnel to avoid these concerns or providing measures to deal with them is critical during the planning process.

The selection of a tunnel alignment should take into consideration site specific constraints such as the presence of contaminated materials, special existing buildings and surface facilities, existing utilities, or the presence of sensitive installations such as historical landmarks, educational institutions, cemetreies, or houses of worship. If certain site constraints cannot be avoided, construction methodologies, and special provisions should be provided. For example, if the presence of contaminated materials near the surface cannot be avoided, a deeper alignment and/or the use of mined excavation (TBM or SEM) would be more suitable than cut and cover method. Similarly, if sensitive facilities exist at the surface and cannot be avoided, special provisions to minimise vibration, and potential surface settlement should be provided in the construction methods.

Risk assessment is an important factor in selecting a tunnel alignment. Construction risks include risks related the construction of the tunnel itself, or related to the impact of the tunnel construction on existing facilities. Some methods of tunnelling are inherently more risky than others or may cause excessive ground movements. Sensitive existing structures may make use of such construction methods in their vicinity undesirable. Similarly, hard spots (rock, for example) beneath parts of a tunnel can also cause undesirable effects and alignment changes may obviate that. Therefore, it is important to conduct risk analysis as early as possible to identify potential risks due to the tunnel alignment and to identify measures to reduce or manage such risks. An example of risk mitigation related to tunnel alignment being close to sensitive surface facilities is to develop and implement a comprehensive instrumentation and monitoring program, and to apply corrective measures if measured movements reach certain thresholds.

Sometimes, modifications in the tunnel structure or configurations would provide benefits for the overall tunnel construction and cost. For example, locating the tunnel ventilations ducts on the side, rather than at the top would reduce the tunnel height, raise the profile of the tunnel and consequently reduce the overall length of the tunnel.

Environmental and Community Issues

Road tunnels are more environmentally friendly than other surface facilities. Traffic congestion would be reduced from the local streets. Air quality would be improved because traffic generated pollutants are captured and disposed of away from the public. Similarly, noise would be reduced and visual aesthetic and land use would be improved. By placing traffic underground, property values would be improved and communities would be less impacted in the long term. Furthermore, tunnels will provide opportunities for land development along and over the tunnel alignment adding real estate properties and potential economical potential development.

In planning for a tunnel, the construction impact on the community and the environment is important and must be addressed. Issues such as impact on traffic, businesses, institutional facilities, sensitive installations, hospitals, utilities, and residences should be addressed. Construction noise, dust, vibration, water quality, aesthetic, and traffic congestion are important issues to be addressed and any potentially adverse impact should be mitigated. For example, a cut-and-cover tunnel requires surface excavation impacting traffic, utilities, and potentially nearby facilities. When completed, it leaves a swath of disturbed surface-level ground that may need landscaping and restoration. In urban situations or close to properties, cut-and-cover tunnels can be disruptive and may cut off access and utilities temporarily. Alternative access and utilities to existing facilities may need to be provided during construction or, alternatively, staged construction to allow access and to maintain the utilities would be required. Sometimes, top-down construction rather than bottom-up construction can help to ameliorate the disruption and reduce its duration. Rigid excavation support systems and ground improvement techniques may be required to minimise potential settlements and lateral ground deformations, and their impact on adjacent structures. When excavation and dewatering are near contaminated ground, special measures may be required to prevent migration of the groundwater contaminated plume into the excavation or adjacent basements. Dust suppression and wheel washing facilities for vehicles leaving the construction site are often used, especially in urban areas.

Similarly, for immersed tunnels the impact on underwater bed level and the water body should be assessed. Dredging will generate bottom disturbance and create solid turbidity or suspension in the water. Excavation methods are available that can limit suspended solids in the water to acceptable levels. Existing fauna and flora and other ecological issues should be investigated to determine whether environmentally and ecologically adverse consequences are likely to ensue. Assessment of the construction on fish migration and spawning periods should be made and measures to deal with them should be developed. The potential impact of construction wetlands should be investigated and mitigated.

On the other hand, using bored tunnelling would reduce the surface impact because generally the excavation takes place at the portal or at a shaft resulting in minimum impact on traffic, air and noise quality, and utility and access disturbance.

Excavation may encounter contaminated soils or ground water. Such soils may need to be processed or disposed in a contained disposal facility, which may also have to be capped to meet the environmental regulations. Provisions would need to address public health and safety and meet regulatory requirements.

Operational Issues

In planning a tunnel, provisions should be made to address the operational and maintenance aspects of the tunnel and its facilities. Issues such as traffic control, ventilation, lighting, life safety systems, equipment maintenance, tunnel cleaning, and the like, should be identified and provisions made for them during the initial planning phases. For example, items requiring more frequent maintenance, such as light fixtures, should be arranged to be accessible with minimal interruption to traffic.

Sustainability

Tunnels by definition are sustainable features. They typically have longer life expectancy than a surface facility (125 versus 75 years). Tunnels also provide opportunities for land development for residential, commercial, or recreational facilities. They enhance the area and potentially increase property values. An example is the "Park on the Lid" in Mercer Island, Seattle, Washington where a park with recreational facilities was developed over I-90. Tunnels also enhance communities connections and adhesion and protect residents and sensitive receptors from traffic pollutants and noise.

General Description of Various Tunnel Types

The principal types and methods of tunnel construction that are in use are:

- Cut-and-cover tunnels are built by excavating a trench, constructing the concrete structure in the trench, and covering it with soil. The tunnels may be constructed in place or by using precast sections
- Bored or mined tunnels, built without excavating the ground surface. These tunnels are usually labelled according to the type of material being excavated. Sometimes a tunnel passes through the boundary between different types of material; this often results in a difficult construction known as mixed face.

- Rock tunnels are excavated through the rock by drilled and blasting, by mechanized excavators in softer rock, or by using rock tunnel boring machines (TBM). In certain conditions, Sequential Excavation Method (SEM) is used.
- Soft ground tunnels are excavated in soil using a shield or pressurized face TBM (principally earth pressure balance or slurry types), or by mining methods, known as either the sequential excavation method (SEM).
- Immersed tunnels, are made from very large precast concrete or concrete-filled steel elements that are fabricated in the dry, floated to the site, placed in a prepared trench below water, and connected to the previous elements, and then covered up with backfill.
- Jacked box tunnels are prefabricated box structures jacked horizontally through the soil using methods to reduce surface friction; jacked tunnels are often used where they are very shallow but the surface must not be disturbed, for example beneath runways or railroads embankments.

Preliminary road tunnel type selection for conceptual study after the route studies can be dictated by the general ground condition.

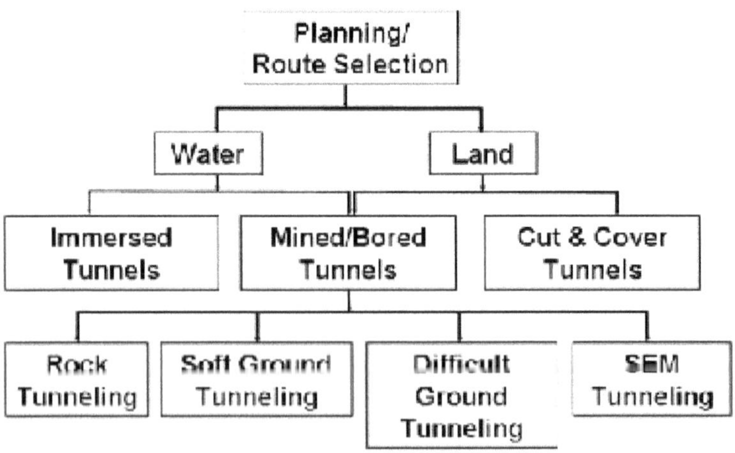

Figure: Preliminary Road Tunnel Type Selection Process

The selection of a tunnel type depends on the geometrical configurations, the ground conditions, the type of crossing, and environmental requirements. For example, an immersed tunnel may be most suitable for crossing a water body, however, environmental and regulatory requirements might make this method very expensive or infeasible. Therefore, it is important to perform the tunnel type study as early as possible in the planning process and select the most suitable tunnel type for the particular project requirements.

Design Process

The basic process used in the design of a road tunnel is:

- Define the functional requirements, including design life and durability requirements;
- Carry out the necessary investigations and analyses of the geologic, geotechnical and geohydrological data
- Conduct environmental, cultural, and institutional studies to assess how they impact the design and construction of the tunnel;
- Perform tunnel type studies to determine the most appropriate method of tunnelling.
- Establish design criteria and perform the design of the various tunnel elements. Appropriate initial and final ground support and lining systems are critical for the tunnel design, considering both ground conditions and the proposed method of construction. Perform the design in Preliminary and Final design phases. Interim reviews should be made if indicated by ongoing design issues.
- Establish tunnel alignment, profile and cross-section
- Determine potential modes of failure, including construction events, unsatisfactory long-term performance, and failure to meet environmental requirements. Obtain any necessary data and analyse these modes of failure;
- Perform risk analysis and identify mitigation measures and implement those measures in the design
- Prepare project documents including construction plans, specifications, schedules, estimates, and geotechnical baseline report (GBR).

Tunnel Cross-Section

The tunnel cross section geometrical configuration must satisfy the required traffic lanes, shoulders or safety walks, suitable spaces for ventilation, lights, traffic control system, fire/life safety systems, etc. The cross section is also dictated by the method of tunnel construction. For example, bored tunnels using TBM will result in circular configuration, while cut and cover construction will result in rectangular configuration. The structural systems will also vary accordingly. The available spaces in a circular cross section can be used to house tunnel systems, such as the ventilation duct or fans, lighting, traffic control systems and signs, close circuit TV, and the like. For rectangular sections the various systems can be placed overhead, invert or adjacent to the traffic lanes if overhead space is limited. It is essential at early design stages to pay attention to detail in laying out the tunnel cross-section to permit easy inspection and maintenance not only of mechanical and electrical equipment, but also of the tunnel structure itself.

The tunnel structural systems depend on the type of tunnel, the geometrical configuration of the cross section, and method of construction. For example, in cut and cover tunnels of rectangular cross section, cast in place concrete is often the selected structural system, while for SEM/NATM tunnel, the structural system could be lattice girders and shotcrete. For soft ground tunnels using TBM, the structural system is often a precast segmental one pass lining. Sometimes, the excavation support system can be used as the final tunnel structural system such as the case in top down construction.

Geometrical Configurations

Groundwater Control

Building a dry tunnel is a primary concerns of the owner, user, and operator alike. A dry tunnel provides a safer and friendlier environment and significantly reduces operation and maintenance costs. Advancements in tunnelling technology in the last few decades in general and in the waterproofing field in particular have facilitated the implementation of strict water infiltration criteria and the ability to build dry tunnels.

Based on criteria obtained from the International Tunnolling Association (ITA), Singapore's Land Transport Authority (LTA), Singapore's Public Utilities Board (PUB), Hong Kong's Mass Transit Rail Corporation (MTRC) and the German Cities Committee, as well as criteria used by various projects in the US and abroad for both highway and transit tunnels (e.g. Washington DC, San Francisco, Atlanta, Boston, Baltimore, Buffalo, Melbourne (Australia), Tyne & Wear (UK) and Antwerp (Belgium), the following ITA ground water infiltration criteria are recommended;

Allowable Infiltration	
Tunnels	≤ 0.002 gal/sq. ft/day
Underground public space	≤ 0.001 gal/sq. ft/day

In addition no dripping or visible leakage from a single location shall be permitted.

Tunnel waterproofing systems are used to prevent groundwater inflow into an underground opening. They consist of a combination of various materials and elements. The design of a waterproofing system is based on the understanding of the ground and geohydrological conditions, geometry and layout of the structure and construction methods to be used. A waterproofing system should always be an integrated system that takes into account intermediate construction stages, final conditions of structures and their ultimate usage including maintenance and operations.

There are two basic types of waterproofing systems: drained (open) and undrained (closed). Various waterproofing materials are available for these systems. Open waterproofing systems allow groundwater inflow into a tunnel drainage system. Typically, the tunnel vault area is equipped with a waterproofing system forming an umbrella-like protection that drains the water seeping towards the cavity around the arch into a drainage system that is located at the bottom of the tunnel sidewalls and in the tunnel invert. The open system is commonly used in rock tunnels where water infiltration rates are low. Groundwater inflow is typically localised to distinct locations such as joints and fractures and the overall permeability is such that a groundwater draw-down in soil layers overlying the rock mass will not be affected. This system is commonly installed between an initial tunnel support (initial lining) and the secondary or final support (permanent lining). The open waterproofing system generally allows for a more economical secondary lining and invert design as the hydrostatic load is greatly reduced or eliminated.

Closed waterproofing systems (closed system), often referred to as tanked systems, extend around the entire tunnel perimetre and aim at excluding the groundwater from flowing into the tunnel drainage system completely. Thus no groundwater drainage is provided. The secondary linings therefore have to be designed for full hydrostatic water pressures. These systems are often applied in permeable soils where groundwater discharge into the tunnels would be significant and would otherwise cause a lowering of the groundwater table and possibly cause surface settlements.

For precast segmental lining, the segments are usually equipped with gaskets to seal the joints between segments and thus provide a watertight tunnel. For cut and cover tunnels under the groundwater table and for immersed tunnels, waterproofing membranes encapsulating the structures are recommended.

Tunnel Portals

Portals and ventilation shafts should be located such that they satisfy environmental and air quality requirements as well as the geometrical configuration of the tunnel. At portals, it may be necessary to extend the dividing wall between traffic travelling in opposite directions to reduce recirculation of pollutants from the exit tunnel into the entry tunnel. If possible, Portals should be oriented to avoid drivers being blinded by the rising or setting sun. Special lighting requirements at the portal are needed to address the "black hole" effect. The portal should be located at a point where the depth of the tunnel is suitably covered. This depends on the type of construction, the crossing configuration, and the geometry of the tunnel. For example, in a cut and cover tunnel, the portal can be as close to the surface as the roof of the tunnel can be placed with sufficient clearance for traffic. On the other hand, in TBM mined tunnels, the portal will be placed at a location where there is sufficient ground cover to start the TBM. In mountain tunnels the portal can be as close to the face of the mountain as practically constructible.

Fire-Life Safety Systems

Figure: Gotthard Tunnel Fire in October 2001 (FHWA 2006)

Safety in the event of a fire is of paramount importance in a tunnel. The catastrophic consequences of the tunnel fires (e.g., the Mont Blanc tunnel, 1999 and the Swiss St. Gotthard tunnel, 2001) not only resulted in loss of life, severe property damages, but also great concerns of the lack of fire-life safety protection in road tunnels. During the Gotthard Tunnel October 2001 fire that claimed 11 deaths, the temperature reportedly reached 1,832 °F (1,000 °C) in few minutes, and thick smoke and combustible product propagated over 1.5 mile (2.5km) within 15 minutes.

For planning purposes, it is important to understand the fire-life safety issues of a road tunnel and consider their impacts on the alignments, tunnel cross section, emergency exits, ventilation provisions, geometrical configuration, right-of-way, and conceptual cost estimates, National Fire Protection Association (NFPA) 502 – Standard for Road Tunnels, Bridges, and Other Limited Access Highways provides the following fire protection and life safety requirements for road tunnels:

- Protection of Structural Elements
- Fire Detection
- Communication Systems
- Traffic Control
- Fire Protection (i.e., standpipe, fire hydrants, water supply, portable fire extinguisher, fixed water-base fire-fighting systems, etc.)
- Tunnel Drainage System
- Emergency Egress
- Electric, and
- Emergency response plan.

In 2005, the FHWA, AASHTO, and the National Cooperative Highway Research Program (NCHRP) sponsored a scanning study of equipment, systems and procedures used in European tunnels. The study concluded with nine (9) recommendations for implementation include conducting research on tunnel emergency management that includes human factors; developing tunnel design criteria that promote optimal driver performance during incidents; developing more effective visual, audible, and tactile signs for escape routes; and using a risk-management approach to tunnel safety inspection and maintenance. Appendix A presents the executive summary of the scan study.

Emergency Egress

Emergency egress for persons using the tunnel to a place of refuge should be provided at regular intervals. Throughout the tunnel, functional, clearly-marked escape routes should be provided for use in an emergency.

Figure: Emergency Exit (FHWA, 2006)

The emergency egress walkways should be a minimum of 3.6 ft wide and should be protected from oncoming traffic. Signage indicating both direction and distance to the nearest escape door should be mounted above the emergency walkways at reasonable intervals (100 to 150 ft) and be visible in an emergency. The emergency escape routes should be provided with adequate lighting level and connected to the emergency power system.

Figure: Emergency Alcove

Where tunnels are provided in twin tubes, cross passages to the adjacent tube can be considered safe haven. The cross passage should be of at least two-hour fire rating construction, should be equipped with self closing fire rated doors that open in both directions or sliding doors, and the cross passages should be located not more than 656 ft (200m) apart. An emergency walkway at least 3.6 feet (1.12 m) wide should be provided on each side of the cross-passageways.

In long tunnels, sometimes breakdown emergency alcoves (local widening) for vehicles are provided. Some European tunnels also provide at intervals an emergency turn-around for vehicles into the adjacent roadway duct which turn-around would normally be closed by doors.

Emergency Ventilation, Lighting and Communication

An emergency ventilation system should be provided to control smoke and to provide fresh air for the evacuation of passengers and for support to the emergency responders. The emergency ventilation system is often the normal ventilation system operated at higher speeds. Emergency ventilation scenarios should be developed and the operation of the fans would be based on the location of the fire and the direction of the tunnel evacuation. The fans should be connected to an emergency power source in case of failure of the primary power.

Emergency tunnel lighting, fire detection, fire lines, and hydrants should be provided. In certain installations, fire suppression measures such as foam or deluge system have been used. The risk of fire spreading through power cable ducts should be eliminated by dividing cable ducts into fireproof sections, placing cables in cast-in ducts, using fireproof cables where applicable, and other preventative measures. Vital installations should be supplied with fire-resistant cables. Materials used should not release toxic or aggressive gases such as chlorine. Water for fire-fighting should be protected against frost. Fire alarm buttons should be provided adjacent to every cross-passage. Emergency services should be able to approach a tunnel fire in safety.

Emergency telephones should be provided in the tunnels and connected to the emergency power supply. When such a telephone is used, the location of the caller should be identified both at the control centre and by a warning light visible to rescuing personnel. Telephones should be provided at cross-passage doors and emergency exits. Communication systems should give the travelling public the possibility of summoning help and receiving instructions, and should ensure coordinated rescue. Systems should raise the alarm quickly and reliably when unusual operating conditions or emergency situations arise.

Radio coverage for police, fire and other emergency services and staff should extend throughout the tunnel. It is necessary for police, fire and emergency services to use their mobile radios within tunnels and cross-passages. Radio systems should not interfere with each other and should be connected to the emergency power supply to communicate with each other. It is also recommended that mobile telephone coverage be provided.

Tunnel Drainage

Good design anticipates drainage needs. Usually sump-pump systems are provided at the portals and at low points. Roadway drainage throughout the tunnel using drain inlets and drainage pipes should be provided. The drainage system should be designed to deal with surface drainage as well as any groundwater infiltration into the tunnel. Other areas of the tunnels, such as ventilation ducts and potential locations for leakage, should have provision for drainage. Accumulation of ice due to inadequate drainage provisions must be avoided for safe passage.

Operational and Financial Planning

Potential Funding Sources and Cash Flow Requirements

Traditionally State, Federal, and Local funds are the main funding sources for road tunnels. However, recently private enterprises and public-private partnership (PPP) are becoming more attractive potential sources for funding road tunnel projects. For example, the Port of Miami Tunnel has been developed using the PPP approach. Various forms of financing have been applied in various locations in the US and the World. Tolls are often levied on users to help repay construction costs, and to pay operating costs, especially when the roads are financed by private sources. In some cases, bond issues have been used to raise funding for the project.

In developing the funding strategy, it is important to consider and secure the cash flow required to complete the project. In assessing the cash flow analysis, escalation to the year of expenditure should be used. Various indices of escalation rates are available. It is recommended that escalation rates comparable to this type of construction and for the area of the project should be used. Factors such as work load in the area, availability of materials, availability of skilled labour, speciality equipment, and the like, should be taken into consideration. Repayment of loans and the cost of the money should be considered. They may continue for a substantial number of years while the operation and maintenance costs of the tunnel also have to be covered.

Conceptual Level Cost Analysis

At the conceptual level, cost analyses are often based upon the costs per unit measurement for a typical section of tunnel. The historical cost data updated for inflation and location is also commonly used as a quick check. However, such data should be used with extreme caution since in most cases, the exact content of such data and any special circumstances are not known. In addition, construction of tunnels is a speciality work and involves a significant labour component. Labour experience and productivity are critical for proper estimating of a tunnel construction cost. Furthermore, the tunnel being a linear structure, its cost is highly dependent on the advance rate of construction, which in turn is dependent on the labour force, the geological conditions, the suitability of equipment, the contractor's means and methods, and the experience of the workers. Since tunnelling is highly dependent on the labour cost, issues such as advance rates, construction schedule, number of shifts, labour union requirements, local regulations such as permissible time of work, environmental factors such as noise and vibrations, and the like should be taken into considerations when construction cost estimates are made. It is recommended, even at the planning phase, to prepare a bottom up construction cost estimate using estimate materials, labour, and equipment. The use of experience from other similar projects in the area is usually done for predicting labour force and the advance rates. At the conceptual level, substantial contingencies may be required at the early stages of a project. As the design advances and the risks identified and dealt with, contingencies would be reduced gradually as the level of detail and design increases. Soft costs such as engineering, program and construction management, insurance, owner cost, third party cost, right of way costs, and the like should be considered.

Project Delivery Methods

Generally, two categories of delivery methods have been used in the past for underground construction, with various levels of success. They are:

- Design-Bid-Build
- Design-Build

The contractual terms of these two delivery methods vary widely. The most common is the fixed price approach, although for tunnelling, the unit prices approach is the most suitable. Other contract terms used include:

- Fixed Price lump sum
- Low bid based on unit prices
- Quality based selection
- Best and Final Offer (BAFO)
- Cost plus fixed fee

The traditional project delivery model is the design-bid-build. In this method, the client finances the project and develops an organisation to deal with project definition, legal, commercial, and land access/ acquisition issues. It appoints a consulting engineer under a professional services contract to act on its behalf to undertake certain design, procurement, construction supervision, and contract administration activities, in return for which the consulting engineer is paid a fee.

The client places construction contracts following a competitive tendering process for a fixed price, with the selection are often based on low bid. This type of contract is simple, straight forward and familiar to public owners. However, in this process the majority of construction risk is passed to the contractor who often uses higher contingency factors to cover the potential construction risks. The client effectively pays the contractor for taking on the risk, irrespective of whether the risk actually transpires.

Whilst this type of contract has its advantages, its shortfalls particularly on large infrastructure projects could be significant. Adversarial relationships between project participants, potential cost overruns, and delays to project schedules are by no means unusual. With the traditional contract forms, there is significant potential for protracted disputes over responsibility for events, to the detriment of the progress of the physical works. The client, its agents, and the contractors are subject to different commercial risks and potentially conflicting commercial objectives.

In a design-build process, the project is awarded to a design-build entity that design and construct the project. The owner's engineer usually prepares bidding documents based on a preliminary-level design identifying the owner's requirements. Contract terms vary from fixed price to unit prices, to cost plus fee. For tunnelling projects, the geotechnical and environmental investigations should be advanced to a higher level of completion to provide better information and understanding of the construction risks. The selected contractor then prepares the final design (usually with consultation with the owner's engineer) and constructs the project. This process is gaining interest among owners of underground facilities in order to reduce the overall time required to complete the project, avoid dealing with disputes over changed conditions, and avoid potential lengthy and costly litigations.

The procurement options of the design-build approach vary based on the project goals and the owners' objectives. Examples of the procurement options include:

- Competitive bid (low price)
- Competitive bid with high responsibility standards (cost and qualifications)
- Competitive bids with alternative proposals
- Price and other factors
- Price after discussion including “Best and Final Offer”
- Quality based selection
- Sole source negotiation

The allocation of risk between the owner and the contractor will have a direct relationship to the contractor contingency as part of the contractor’s bid. Therefore, it is important to identify a risk sharing mechanism that is fair and equitable and that will result in a reasonable contingency by the contractor and sufficient reserve fund to be provided by the owner to address unforeseen conditions. For example unforeseen conditions due to changes in the anticipated ground conditions are paid for by the owner if certain tests are met, while the means and methods are generally the contractor’s responsibility and his inability to perform under prescribed conditions are risks to be absorbed by the contractor. With proper contracting form and equitable allocation of risks between the owner and the contractor, the contractor contingency, which is part of its bid price, will be reduced. Similarly, the owner’s reserve fund will be used only if certain conditions are encountered, resulting in an overall lesser cost to the owner.

Design-build has the advantage that the design can be tailored to fit the requirements of the contractor‘s means and methods since both, the designer and the constructor work through one contract. This can be particularly useful when some of the unknown risks are included in the contractor’s price without major penalties that could occur if the design is inadequate. Risk sharing is especially useful if anticipated conditions can be defined within certain limits and the client takes the risk if the limits are exceeded. Examples of conditions that might not be expected include soil behaviour, the hardness of rock, flood levels, extreme winds and currents. Considerable use is currently made of Geotechnical Baseline Reports to define anticipated ground conditions in this way.

Most claims in tunnel construction are related to unforeseen ground conditions. Therefore, the underground construction industry in the US tried to provide a viable trigger by means of the Differing Site Condition (DSC) clause, culminating in the use of the Geotechnical

Baseline Report (GBR) and Geotechnical Data Report (GDR). It is important from a risk-sharing perspective that the contractual language in the DSC and the GBR are complementary.

It is important to establish a selection process by which only qualified contractors can bid on tunnelling projects, with fair contracts that would allocate risks equitably between the owner and the contractor, in order to have safe, on time, and high quality underground projects at fair costs.

Operation and Maintenance Cost Planning

Operations are divided into three main areas, traffic and systems control, toll facility (if any), and emergency services, not all of which may be provided for any particular tunnel. The staff needed in these areas would vary according to the size of the facility, the location, and the needs. For 24-hour operation, staff would be needed for three shifts and weekends; weekend and night shifts would require sufficient staff to deal with traffic and emergency situations.

The day-to-day maintenance of the tunnel generally requires a dedicated operating unit. Tunnel cleaning and roadway maintenance are important and essential for safe operation of the tunnel. Special tunnel cleaning equipment are usually employed. Mechanical, electrical, communication, ventilation, monitoring, and control equipment for the tunnel must be kept operational and in good working order, since faulty equipment could compromise public safety. Regular maintenance and 24-hour monitoring is essential, since failure of equipment such as ventilation, lights and pumps is unacceptable and must be corrected immediately. Furthermore, vehicle breakdowns and fires in the tunnel need immediate response.

Generally most work can be carried out during normal working hours including mechanical and electrical repair, traffic control, and the like. However, when the maintenance work involves traffic lane closure, such as changing lighting fixtures, roadway repairs, and tunnel washing, partial or full closure of the tunnel may be required. This is usually done at night or weekends.

Detailed discussions for the operation and maintenance issues are beyond the scope of this manual.

Risk Analysis and Management

Risk analysis and management is essential for any underground project. A risk register should be established as early as possible in the project development. The risk register would identify potential risks, their probability of occurrence and their consequences. A risk management plan should be established to deal with the various risks either by eliminating them or reducing their consequences by planning, design, or by operational provisions. For risks that cannot be mitigated, provisions must be made to reduce their consequences and to manage them. An integrated risk management plan should be regularly updated to identify all risks associated with the design, execution and completion of the tunnel. The plan should include all reasonable risks associated with design, procurement and construction. It should also include risks related to health and safety, the public and to the environment.

Major risk categories include construction failures, public impact, schedule delay, environmental commitments, failure of the intended operation and maintenance, technological challenges, unforeseen geotechnical conditions, and cost escalation.

Geometrical Configuration

Provides general geometrical requirements for planning and design of road tunnels. The topics consist of the following: horizontal and vertical alignments; clearance envelopes; and cross section elements. Geometrical requirements for the tunnel approaches and portals are also provided. In addition to the requirements addressed herein, the geometrical configurations of a road tunnel are also governed by its functionality and locality, as well as the subsurface conditions and its construction method (i.e. cut-and-cover, mined and bored, immersed, etc.). It often takes several iterative processes from planning, environmental study, configuration, and preliminary investigation and design to eventually finalize the optimum alignment and cross section layout.

Introduction

As defined by the American Association of State Highway and Transportation Officials (AASHTO) Technical Committee for Tunnels (T-20), road tunnels are defined as enclosed roadways with vehicle access that is restricted to portals regardless of type of structure or method of construction. Road tunnels following this definition exclude enclosed roadway created by air-rights structures such as highway bridges, railroad bridges or other bridges.

In addition to the general roadway requirements, road tunnels also require special considerations including lighting, ventilation, fire protection systems, and emergency egress capacity. These considerations often impose additional geometrical requirement as discussed in the following sections.

Design Standards

Road tunnels discussed in this Manual cover all roadways including freeways, arterials, collectors, and local roads and streets in urban and rural locations following the functional classifications from FHWA publication "Highway Functional Classification: Concepts, Criteria, and Procedures." AASHTO's "Green Book" - A Policy on Geometric Design of Highways and Streets, which is adopted by Federal agencies, States, and most local highway agencies, provides the general design considerations used for road tunnels from the standpoint of service level, and suggests the requirements for road tunnels which should not differ materially from those used for grade separation structures. The Green Book (AASHTO, 2004) also provides general information and recommendations about cross section elements and other requirements specifically for road tunnels.

Additional criteria may include:

- AASHTO A Policy on Design Standards - Interstate System
- Standards issued by the state or states in which the tunnel is situated
- Local authority standards, where these are applicable
- National and local standards of the country where the international crossing tunnel is located

Although the geometrical requirements for roadway alignment, profile and for vertical and horizontal clearances in the above design standards generally apply to road tunnels, amid the high costs of tunnelling and restricted right-of-way, minimum requirements are typically applied to planning and design of road tunnels to minimise the overall size of the tunnel yet maintain a safe operation through the tunnel. To ensure roadway safety, the geometrical design must evaluate design speed, lane and shoulder width, tunnel width, horizontal and vertical alignments, grade, stopping sight distance, cross slope, superelevation, and horizontal and vertical clearances, on a case by case basis.

In addition to the above highway design standards, geometrical design for road tunnels must consider tunnel systems such as fire life safety elements, ventilation, lighting, traffic control, fire detection and protection, communication, etc... Therefore, planning and design of the alignment and cross section of a road tunnel must also comply with National Fire Protection Association (NFPA) 502 – Standard for Road Tunnels, Bridges, and Other Limited Access Highways.

The recommendations in this Manual are provided as a guide for the engineer to exercise sound judgment in applying standards to the geometrical design of tunnels and generally base on 5th Edition (2004) of the Green Book and 2008 Edition of NFPA 502. The design standards used for a road tunnel project should equal or exceed the minimum given below in this Manual to the maximum extent feasible, taking into account costs, traffic volumes, safety requirements, right of way, socioeconomic and environmental impacts, without compromising safety considerations. The readers should always check with the latest requirements from the above references.

Horizontal and Vertical Alignments

Maximum Grades: Road tunnel grades should be evaluated on the basis of driver comfort while striving to reach a point of economic balance between construction costs and operating and maintenance expenses.

Maximum effective grades in main roadway tunnels preferably should not exceed 4%; although grades up to 6% have been used where necessary. Long or steep uphill grades may result in a need for climbing lanes for heavy vehicles. However, for economic and ventilation reasons, climbing lanes should be avoided within tunnels; the addition of a climbing lane part-way through a tunnel may also complicate construction considerably, particularly in a bored tunnel.

Horizontal and Vertical Curves

Horizontal and vertical curves shall satisfy Green Book's geometrical requirements. The horizontal alignment for a road tunnel should be as short as practical and maintain as much of the tunnel length on tangent as possible, which will limit the numbers of curves, minimise the length and improve operating efficiency. However, slight curves may be required to accommodate ventilation/access shafts location, portal locations, construction staging areas, and other ancillary facilities. A slight horizontal curve at the exit of the tunnel may be required to allow drivers to adjust gradually to the brightness outside the tunnel.

When horizontal curves are needed, the minimum acceptable horizontal radii should consider traffic speed, sight distances, and the superelevation provided. In general, for planning purpose, the curve radii should be as large as possible and no less than 850 to 1000-ft radius. A tighter curve may be considered at the detailed design stage based on the selected tunnelling method.

Super elevation rate, which is the rise in the roadway surface elevation from the inside to the outside edge of the road, should preferably lie in the range 1% to 6%.

When chorded construction is used for walls where alignments are curved, chord lengths should not exceed 25 feet (7.6 m) for radii below 2,500 feet (762 m), and 50 feet (15 m) elsewhere.

Sight and Braking Distance Requirements

Sight and braking distance requirements cannot be relaxed in tunnels. On horizontal and vertical curves, it may be necessary to widen the tunnel locally to meet these requirements by providing a "sight shelf." When designing a tunnel with extreme curvature, sight distance should be carefully examined, otherwise it may result in limited stopping sight distance.

Other Considerations

Road tunnels with more than one traffic tube should be designed so that in the event that one tube is shut down, traffic can be carried in the other. For reasons of safety, it is not recommended that tunnels be constructed for bi-directional traffic; however, they should be designed to be capable of handling bi-directional traffic during maintenance work, which should be carried out at times of low traffic volume such as at night or weekends. When operating in a bi-directional mode, appropriate signage must be provided. In addition, suitable cross-over areas are required, usually provided outside the tunnel entrances, and the ventilation system and signage must be designed to handle bi-directional traffic.

For bored and mined tunnels, it is probable that separate tunnels are constructed for traffic in each direction. For cut-and-cover, jacked and immersed tunnels, it is preferable for the traffic tubes for the two directions to be constructed within a single structure so that emergency egress by vehicle occupants into a neighbouring traffic tube can be provided easily. Note that NFPA 502-2008 requires that the two tubes be divided by a minimum of 2-hour fire rated construction in order to consider cross-passageways between the tunnels to be utilised in lieu of emergency egress.

In addition to structural requirements, inundation of the tunnel by floods, surges, tides and waves, or combinations thereof resulting from storms must be prevented. The height and shape of walls surrounding tunnel entrances, the elevation of access road surfaces and any entrances, accesses and holes must be designed such that entry of water is prevented. It is recommended that water level with the probability of being exceeded no more than 0.005 times in any one year (the 500-year flood level) be used as the design water level.

Travel Clearance

Clearance diagram of all potential vehicles traversing the tunnel shall be established using dynamic vehicle envelopes which consider not just the maximum allowable static envelope, but also other dynamic factors such as bouncing, suspension failure, overhang on curves, lateral motion, resurfacing, etc. The clearance diagram should take into consideration potential future vehicle heights, vehicle mounting on curbs, construction tolerances, and any potential ground and structure settlement. Ventilation equipment, lighting, guide signs, and other equipment should not encroach within the clearance diagram.

Vertical clearance should be selected as economical as possible consistent with the vehicle size. The 5th Edition of AASHTO Green Book (2004) recommends that the minimum vertical clearance to be 16 feet (4.9 m) for highways and 14 feet (4.3 m) for other roads and streets. Note that the minimum clear height should not be less than the maximum height of load that is legal in a particular state. The minimum clearance diagram for a two-lane tunnel which indicates the minimum horizontal curb-to-curb and wall-to-wall clearances to be no less than 24 ft (7.2-m) and 30 ft (9-m), respectively. The curb-to-curb (including shoulders) clearance is also required to be 2 ft (0.6m) greater than the lane widths of the approach travelled way. Therefore, for an approach structure with two standard 12-ft lane widths, the minimum horizontal curb-to-curb clearance for the connecting two-lane tunnel should be no less than 26 ft (7.8 m).

The vertical clearance shall also take into consideration for future resurfacing of the roadway. Although it is recommended to resurface roadways in tunnels only after the previous surface has been removed, it is prudent to provide limited allowances for resurfacing once without removal of the old pavement. Consideration should also be given for potential truck mounting on the barrier in the tunnel or on low sidewalk and measures shall be used to prevent such mounting from damaging the tunnel ceiling or tunnel system components mounted on the ceiling or the walls.

Tunnel ventilation ducts, if required, can be provided above or below the traffic lanes, or to the sides of them. Where clearances to the outside of the tunnel at a particular location are such that by moving ventilation from overhead to the sides can reduce the tunnel gradients or reduce its length, such an option should be considered.

Over-height warning signals and diverging routes should be provided before traffic can reach the tunnel entrances. The designated traffic clearance should be provided throughout the approaches to the tunnel.

Cross Section Elements

Typical Cross Section Elements

Although many road tunnels appear rectangular from inside bordered by the walls, ceiling and pavement, the actual tunnel shapes may not be rectangular. As described, there are generally three typical shapes of tunnels – circular, rectangular, and horseshoe/ curvelinear..

Figure: A Typical Horseshoe Section for a Two-lane Tunnel (Glenwood Canyon, Colorado)

A road tunnel cross section must be able to accommodate the horizontal and vertical traffic clearances, as well as the other required elements. The typical cross section elements include:

- Travel lanes
- Shoulders
- Sidewalks/Curbs
- Tunnel drainage
- Tunnel ventilation
- Tunnel lighting
- Tunnel utilities and power
- Water supply pipes for firefighting

- Cabinets for hose reels and fire extinguishers
- Signals and signs above roadway lanes
- CCTV surveillance cameras
- Emergency telephones
- Communication antennae/equipment
- Monitoring equipment of noxious emissions and visibility
- Emergency egress illuminated signs at low level (so that they are visible in case of a fire or smoke condition+)

Additional elements may be needed under certain design requirements and should be taken into consideration when developing the tunnel geometrical configuration. The requirements for travel lane and shoulder width, sidewalks/emergency egress, drainage, ventilation, lighting, and traffic control are discussed in the following sections. Other elements cited above are required for fire and safety protection for tunnels longer than 1000 ft (300m) or 800 ft (240m) long if the maximum distance from any point within the tunnel to a point of safety exceeds 400 ft (120m) (NFPA, the latest). Fire and safety protection requirements are not within the scope of this manual. Refer to Appendix A, and the latest NFPA 502 Standard for requirement for fire and safety protection elements.

Travel Lane and Shoulder

As discussed previously, for planning and design purposes, each lane width within a road tunnel should be no less than 12 feet (3.6 m) as recommended in the 5th Edition of Green Book (AASHTO, 2004).

Although the Green Book states that it is preferable to carry the full left- and right-shoulder widths of the approach freeway through the tunnel, it also recognises that the cost of providing full shoulder widths may be prohibitive. Reduction of shoulder width in road tunnels is usual. In certain situations narrow shoulders are provided on one or both sides. Sometimes shoulders are eliminated completely and replaced by barriers. Based on a study conducted by World Road Association (PIARC) and published a report entitled "Cross Section Geometry in Unidirectional Road Tunnels" 2001; shoulder widths vary from country to country and they range from 0 to 2.75 m (9 ft). They are generally in the range of 1 m (3.3 ft). It is suggested for unidirectional road tunnels that the right shoulder be at 4 ft (1-2 m) and left shoulder at least 2 ft (0.6 m).

A minimum requirement for a shoulder in a tunnel, except it requires that a minimum 2 feet (0.6m) be added to the travel lane width of the approach structure. The Green Book also recommends that the determination of the width of shoulders be established on an in-depth analysis of all aspects involved. Where it is not realistic (for economic or constructability considerations) to provide shoulders at all in a tunnel, travel delays may occur when vehicle(s) become inoperative during periods of heavy traffic. In long tunnels, emergency alcoves are sometimes provided to accommodate disabled vehicles.

To prevent errant vehicles from hitting the walls of the tunnel, a deflecting concrete barrier with a sloping or partially sloping face is commonly used. The height of the barrier should not be so great that it is perceived by drivers of low vehicles to be narrowing the width to the wall nor should it be too low to allow vehicles to mount it. A barrier of 3.3 ft (1 m) is common. A reduced shoulder width from a travelled way to the face of the adjacent barrier ranging between 2 and 4 feet (between 0.6 and 1.2 m) has been found to be acceptable.

A typical tunnel roadway section including and two standard 12 ft lane widths and two reduced shoulder widths. Refer to Section 2.4.3 for the requirements for the barriers when used as the raised sidewalks or emergency egress walkways.

Sidewalks/Emergency Egress Walkway

Although pedestrians are typically not permitted in road tunnels, sidewalks are required in road tunnels to provide emergency egress and access by maintenance personnel. The 5th Edition of Green Book recommends that raised sidewalks or curbs with a width of 2.5 ft (0.7 m) or wider beyond the shoulder area are desirable to be used as an emergency egress, and that a raised barrier to prevent the overhang of vehicles from damaging the wall finish or the tunnel lighting fixtures be provided.

In addition, NFPA 502 requires an emergency egress walkway within the cross-passageways be of a minimum clear width of 3.6 ft (1.12 m).

Tunnel Drainage Requirements

Road tunnels must be equipped with a drainage system consisting of pipes, channels, sump/pump, oil/water separators and control systems for the safe and reliable collection, storage, separation and disposal of liquid/ effluent from the tunnels that might otherwise collect. Drainage must be provided in tunnels to deal with surface water as well as water leakage. However, drainage lines and sump-pumps should be sized to accommodate water intrusion and/or fire fighting requirements. They should be designed so that fire would not spread through the drainage system into adjacent tubes by isolating them. For the safety reason, PVC, fibreglass pipe, or other combustible materials should not be used.

Sumps should be provided with traps to collect and remove solids. Sand traps should be provided, as well as oil and fuel separators. It may be assumed in sizing sumps that fires and storms do not happen simultaneously. Sumps and pumps should be located at low points of a tunnel and at portals to handle water that might otherwise flow into the tunnel. Sumps should be sized to match the duty cycle of the discharge pumps such that inflow does not cause sump capacity to be exceeded. Sumps should be designed to be capable of being cleaned regularly.

Ventilation Requirements

The ventilation system of a tunnel operates to maintain acceptable air quality levels for short-term exposure within the tunnel. The design may be driven either by fire/safety considerations or by air quality; which one governs depends upon many factors including traffic, size and length of the tunnel, and any special features such as underground interchanges.

Ventilation requirements in a highway tunnel are determined using two primary criteria, the handling of noxious emissions from vehicles using the tunnel and the handling of smoke during a fire. Computational fluid dynamics (CFD) analyses are often used to establish an appropriate design for the ventilation under fire conditions. An air quality analysis should also be conducted to determine whether air quality might govern the design. Air quality monitoring points in the tunnel should be provided and the ventilation should be adjusted based on the traffic volume to accommodate the required air quality.

Environmental impacts and air quality may affect the locations of ventilation structures/buildings, shafts and portals. Analyses should take into account current and future development, ground levels, the heights and distances of sensitive receptors near such locations and the locations of operable windows and terraces of adjacent buildings to minimise impacts. Ventilation buildings have also been located below grade and exhaust stacks hidden within other structures.

The two main ventilation system options used for tunnels are longitudinal ventilation and transverse ventilation. A longitudinal ventilation system introduces air into, or removes air from a road tunnel, with the longitudinal flow of traffic, at a limited number of points such as a ventilation shaft or a portal. It can be sub-classified as either using a jet fan system or a central fan system with a high-velocity (Saccardo) nozzle.

The use of jet fan based longitudinal system was approved by the FHWA in 1995 based on the results of the Memorial Tunnel Fire Ventilation Test Program (NCHRP, 2006). Generally, it includes a series of axial, high-velocity jet fans mounted at the ceiling level of the road tunnel to induce a longitudinal air-flow through the length of the tunnel as shown.

Figure: Ventilation System with Jet Fans at Cumberland Gap Tunnel

A transverse ventilation system can be either a full or semi-full transverse type. With full transverse ventilation, air supply ducts are located above, below or to the side of the traffic tube and inject fresh air into the tunnel at regular intervals. Exhaust ducts are located above or to the side of the traffic tube and remove air and contaminants. With semi-transverse ventilation, the supply duct is eliminated with its "duties" taken over by the traffic opening. When supply or exhaust ducts are used, the flow is generated by fans grouped together in ventilation buildings. Local noise standards generally would require noise attenuators at the fans or nozzles.

Selection of the appropriate ventilation system obviously has a profound impact on the tunnel alignment, layout, and cross section design.

Lighting Requirements

Lighting in tunnels assists the driver in identifying hazards or disabled vehicles within the tunnel while at a sufficient distance to safely react or stop. High light levels (Portal light zone) are usually required at the beginning of the tunnel during the daytime to compensate for the "Black Hole Effect" that occurs by the tunnel structure shadowing the roadway as shown. These high light levels will be used only during daytime.

Tunnel light fixtures are usually located in the ceiling, or mounted on the walls near the ceiling. Tunnel lighting methods and guidelines are not within the scope of this manual. However, the location, size, type, and number of light fixtures impact the geometrical requirements of the tunnel and should be taken into consideration.

Figure 2-7 "Black Hole" (Left) and Proper Lighting (Right)

The tunnel lighting documents issued by the IESNA (ANSI/IESNA RP-22 Recommended Practice for Tunnel Lighting) and the CIE (CIE-88 Guide for the Lighting of Road Tunnels and Underpasses) offer comprehensive approaches to tunnel lighting. The AASHTO Roadway Lighting Design Guide provides some recommendations for road tunnels as well.

For improved safety during a fire, it is suggested that strobe lights be placed to identify exit routes. If used they should be placed around exit doors, especially at lower levels which might then be under the smoke level.

The strobe lights would be activated only during tunnel fires. Emergency lighting in tunnels including wiring methods and other requirements are included in NFPA 502 "Standard for Road Tunnels, Bridges and Other Limited Access Highways", PIARC "Fire and Smoke Control in Road Tunnels", and in the findings of the 2005 FHWA/AASHTO European Scan Tour.

Traffic Control Requirements

The latest NFPA 502 Standard mandates that tunnels 300 ft (90 m) in length should be provided with a means to stop traffic approaching portals (external). In addition, the NFPA 502 also specifies that traffic control means within the tunnel 800 feet (240 m) in length are required.

These should include lane control signals, over-height warning signals, changeable message signs (CMS), etc. Traffic control may be required to close and open lanes for maintenance and handling accidents, and for monitoring of vehicles carrying prohibited materials.

Incident control systems linked to CCTV cameras should be installed. It is recommended that 100% coverage of the tunnel with CCTV be provided. Traffic control requirements should be taken into consideration when developing the cross sectional geometry.

Portals and Approach

Tunnel portals may require special design considerations. Portal sites need to be located in stable ground with sufficient space. Orientation of the portals should avoid if possible direct East and West to avoid blinding sunlight.

Ameliorating measures should be taken where drivers might otherwise be blinded by the rising or setting sun. Intermittent cross members are sometimes provided across the approach structure above the traffic lanes as an amelioration measure. A central dividing wall sometime is extended some distance out from the portal to prevent recirculation of polluted air, i.e. vented polluted air from one traffic duct is prevented from entering an adjacent duct as "clean" air.

Tunnels with a high traffic volume and long tunnels should be equipped with emergency vehicles at each end with potential access to all traffic tubes. Wrecker trucks should be capable of pushing a disabled vehicle as well as the more traditional method of towing. These vehicles should preferably be equipped with some fire-fighting equipment, the extent of which depending upon the distance to the nearest fire department. At least, they should carry dry chemical fire extinguishers.

If the tunnel is in a remote rural area where responses of nearby fire companies and emergency squads are not available in a timely matter, a larger portal structure as shown, may be required to host the operation control centre, as well as the fire-fighting and emergency-responding personnel, equipment and vehicles.

In determining portal locations and where to end the approach structure and retaining walls, protection should be provided against flooding resulting from high water levels near bodies of water and tributary watercourses, or from storm runoff.

The height of the portal end wall and the approach retaining walls should be set to a level at least 2 ft (0.6m) higher than the design flood level. Alternatively a flood gate can be provided. Adequate provision should be made for immediate and effective removal of water from rainfall, drainage, groundwater seepage, or any other source. Portal cross drain and sump-pump should be provided.

Chapter 3

Geotechnical Investigations

To successfully plan, design and construct a road tunnel project requires various types of investigative techniques to obtain a broad spectrum of pertinent topographic, geologic, subsurface, geo-hydrological, and structure information and data. Although most of the techniques and procedures are similar to those applied for roadway and bridge projects, the specific scope, objectives and focuses of the investigations are considerably different for tunnel and underground projects, and can vary significantly with subsurface conditions and tunnelling methods.

A geotechnical investigation program for a tunnel project must use appropriate means and methods to obtain necessary characteristics and properties as basis for planning, design and construction of the tunnel and related underground facilities, to identify the potential construction risks, and to establish realistic cost estimate and schedule. The extent of the investigation should be consistent with the project scope (i.e., location, size, and budget), the project objectives (i.e., risk tolerance, long-term performance), and the project constraints (i.e., geometry, constructability, third-party impacts, aesthetics, and environmental impact). It is important that the involved parties have a common understanding of the geotechnical basis for design, and that they are aware of the inevitable risk of not being able to completely define existing subsurface conditions or to fully predict ground behaviour during construction.

Generally, an investigation program for planning and design of a road tunnel project may include the following components:

- Existing Information Collection and Study
- Surveys and Site Reconnaissance
- Geologic Mapping
- Subsurface Investigations

- Environmental Studies
- Seismicity
- Geospatial Data Management

It is beyond the scope of this manual to discuss each of the above components in details. The readers are encouraged to review the FHWA and AASHTO. Similar investigations and monitoring are often needed during and after the construction to ensure the problems that occurred during construction are rectified or compensated, and short term impacts are reversed. Geotechnical investigations after construction are not discussed specifically in this Chapter.

Phasing of Geotechnical Investigations

Amid the higher cost of a complete geotechnical investigation program for a road tunnel projects (typically about 3% to 5% of construction cost), it is more efficient to perform geotechnical investigations in phases to focus the effort in the areas and depths that matter. Especially for a road tunnel through mountainous terrain or below water body, the high cost, lengthy duration, limited access, and limited coverage of field investigations may demand that investigations be carried out in several phases to obtain the information necessary at each stage of the project in a more cost-efficient manner.

Furthermore, it is not uncommon to take several decades for a road tunnel project to be conceptualized, developed, designed, and eventually constructed. Typical stages of a road tunnel project from conception to completion are:

- Planning
- Feasibility Study
- Corridor and Alignment Alternative Study
- Environmental Impact Studies (EIS) and Conceptual Design
- Preliminary Design
- Final Design
- Construction

Throughout the project development, the final alignment and profile may often deviate from those originally anticipated. Phasing of the geotechnical investigations provides an economical and rational approach for adjusting to these anticipated changes to the project.

The early investigations for planning and feasibility studies can be confined to information studies and preliminary reconnaissance.

Geological mapping and minimum subsurface investigations are typically required for EIS, alternative studies and conceptual design. EIS studies may also include limited topographical and environmental investigations to identify potential "fatal flaws" that might stop the project at a later date. A substantial portion of the geotechnical investigation effort should go into the Preliminary Design Phase to refine the tunnel alignment and profile once the general corridor is selected, and to provide the detailed information needed for design. As the final design progresses, additional test borings might be required for fuller coverage of the final alignment and for selected shaft and portal locations. Lastly, depending on the tunnelling method selected, additional investigations may be required to confirm design assumptions, or to provide information for contractor design of temporary works.

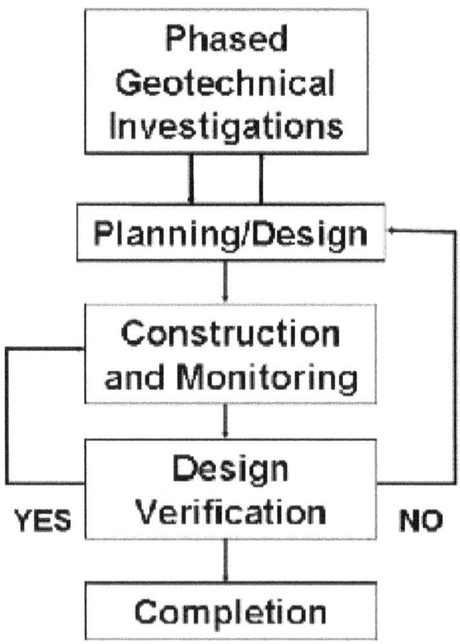

Figure: Phased Geotechnical Investigations with Project Development Process

This Chapter discusses the subsurface investigation techniques typically used for planning, design and construction of road tunnels. Additional information on this subject is available from FHWA Geotechnical Engineering Circular No. 5 (FHWA, 2002a), FHWA Reference Manual for Subsurface Investigations – Geotechnical Site

Characterisation (FHWA, 2002b), FHWA Reference Manual for Rock Slopes (FHWA, 1999), and AASHTO Manual on Subsurface Investigations (AASHTO, 1988).

Information Study

Collection and Review of Available Information

The first phase of an investigation program for a road tunnel project starts with collection and review of available information to develop an overall understanding of the site conditions and constraints at little cost. Existing data can help identify existing conditions and features that may impact the design and construction of the proposed tunnel, and can guide in planning the scope and details of the subsurface investigation program to address these issues.

Published topographical, hydrological, geological, geotechnical, environmental, zoning, and other information should be collected, organised and evaluated. In areas where seismic condition may govern or influence the project, historical seismic records are used to assess earthquake hazards. Records of landslides caused by earthquakes, documented by the USGS and some State Transportation Departments, can be useful to avoid locating tunnel portals and shafts at these potentially unstable areas.

In addition, case histories of underground works in the region are sometimes available from existing highway, railroad and water tunnels. Other local sources of information may include nearby quarries, mines, or water wells. University publications may also provide useful information.

Today, existing data are often available electronically, making them easier to access and manage. Most of the existing information such as aerial photos, topographical maps, etc. can be obtained in GIS format at low or no cost. Several state agencies are developing geotechnical management systems (GMS) to store historical drilling, sampling, and laboratory test data for locations in their states. An integrated project geo-referenced (geospatial) data management system will soon become essential from the initiation of the project through construction to store and manage these extensive data instead of paper records. Such an electronic data management system after the project completion will continue to be beneficial for operation and maintenance purposes.

Topographical Data

Topographic maps and aerial photographs that today can be easily and economically obtained, are useful in showing terrain and geologic features (i.e., faults, drainage channels, sinkholes, etc.). When overlapped with published geological maps they can often, by interpretation, show geologic structures. Aerial photographs taken on different dates may reveal the site history in terms of earthwork, erosion and scouring, past construction, etc.

U.S. Geological Survey (USGS) topographic maps (1:24,000 series with 10 ft or 20 ft contours) may be used for preliminary route selection. However, when the project corridor has been defined, new aerial photography should be obtained and photogrammetric maps should be prepared to facilitate portal and shaft design, site access, right-of-way, drainage, depth of cover, geologic interpretation and other studies.

Topographic Surveys

As alternatives are eliminated, detailed topographic maps, plans and profiles must be developed to establish primary control for final design and construction based on a high order horizontal and vertical control field survey. On a road tunnel system, centreline of the roadway and centreline of tunnel are normally not identical because of clearance requirements for walkways and emergency passages. A tunnel centreline developed during design should be composed of tangent, circular, and transition spiral sections that approximate the complex theoretical tunnel centreline within a specified tolerance (0.25 in.). This centreline should be incorporated into the contract drawings of the tunnel contract, and all tunnel control should be based on this centreline. During construction, survey work is necessary for transfer of line and grade from surface to tunnel monuments, tunnel alignment control, locating and monitoring geotechnical instrumentation (particularly in urban areas), as-built surveys, etc. Accurate topographic mapping is also required to support surface geology mapping and the layout of exploratory borings, whether existing or performed for the project. The principal survey techniques include:

- Conventional Survey
- Global Positioning System (GPS)
- Electronic Distance Measuring (EDM) with Total Stations.
- Remote Sensing
- Laser Scanning

The state-of-art surveying techniques are discussed briefly below. Note that the accuracies and operation procedures of these techniques improve with time so the readers should seek out up-to-date information when applying these techniques for underground projects.

Global Positioning System (GPS) utilises the signal transit time from ground station to satellites to determine the relative position of monuments in a control network. GPS surveying is able to coordinate widely spaced control monuments for long range surveys, as well as shorter range surveys. The accuracy of GPS measurement is dependent upon the number of satellites observed, configuration of the satellite group observed, elapsed time of observation, quality of transmission, type of GPS receiver, and other factors including network design and techniques used to process data. The drawback for GPS survey is its limitation in areas where the GPS antenna cannot establish contact with the satellites via direct line of sight, such as within tunnels, downtown locations, forested areas, etc.

Electronic Distance Measuring (EDM) utilises a digital theodolite with electronic microprocessors, called a "total station" instrument, which determines the distance to a remote prism target by measuring the time required for a laser or infrared light to be reflected back from the target. EDM can be used for accurate surveys of distant surfaces that would be difficult or impractical to monitor by conventional survey techniques. EDM can be used for common surveying applications, but is particularly useful for economically monitoring displacement and settlement with time, such as monitoring the displacement and settlement of an existing structure during tunnelling operations.

Remote Sensing can effectively identify terrain conditions, geologic formations, escarpments and surface reflection of faults, buried stream beds, site access conditions and general soil and rock formations. Remote sensing data can be easily obtained from satellites (i.e. LANDSAT images from NASA), and aerial photographs, including infrared and radar imagery, from the USGS or state geologists, U.S. Corps of Engineers, and commercial aerial mapping service organisations. State DOT aerial photographs, used for right-of-way surveys and road and bridge alignments, may also be available.

Laser Scanning utilises laser technology to create 3D digital images of surfaces. Laser scanning equipment can establish x, y and z coordinates of more than one thousand points per second, at a resolution of about 0.25 inch over a distance of more than 150 feet.

Laser scanning can be used to quickly scan and digitally record existing slopes to determine the geometry of visible features, and any changes with time. These data may be useful in interpreting geologic mapping data, for assessing stability of existing slopes, or obtaining as-built geometry for portal excavations. In tunnels, laser scanning can efficiently create cross sections at very close spacing to document conditions within existing tunnels, verify geometry and provide as-built sections for newly constructed tunnels, and to monitor tunnel deformations with time.

Hydrographical Surveys

Hydrographic surveys are required for subaqueous tunnels including immersed tunnel, shallow bored tunnel, jacked box tunnel, and cofferdam cut-and-cover river crossings to determine bottom topography of the water body, together with water flow direction and velocity, range in water level, and potential scour depth. In planning the hydrographic survey, an investigation should be made to determine the existence and location of submarine pipelines, cables, natural and sunken obstructions, rip rap, etc. that may impact design or construction of the immersed tunnel or cofferdam cut-and-cover tunnel. Additional surveys such as magnetometre, seismic sub-bottom scanning, electromagnetic survey, side scan sonar, etc., may be required to detect and locate these features. These additional surveys may be done simultaneously or sequentially with the basic hydrographic survey. Data generated from the hydrographic survey should be based on the same horizontal coordinate system as the project control surveys, and should be compatible with the project GIS database. The vertical datum selected for the hydrographic survey should be based on the primary monument elevations, expressed in terms of National Geodetic Vertical Datum of 1929 (NGVD), Mean Lower Low Water Datum, or other established project datum.

Utility Surveys

Utility information is required, especially in the urban areas, to determine the type and extent of utility protection, relocation or reconstruction needed. This information is obtained from surveys commissioned for the project, and from existing utility maps normally available from the owners of the utilities (utility companies, municipalities, utility districts, etc.). Utility surveys are performed to collect new data, corroborate existing data, and composite all data in maps and reports that will be provided to the tunnel designer. The requirement for utility information varies with tunnelling methods and site conditions. Cut-and-cover tunnel and shallow soft ground tunnel constructions, particularly in urban areas, extensively impacts overlying and adjacent utilities. Gas, steam, water, sewerage, storm water, electrical, telephone, fibre optic and other utility mains and distribution systems may require excavation, rerouting, strengthening, reconstruction and/or temporary support, and may also require monitoring during construction.

The existing utility maps are mostly for informational purposes, and generally do not contain any warranty that the utility features shown actually exist, that they are in the specific location shown on the map, or that there may be additional features that are not shown. In general, surface features such as manholes and vaults tend to be reasonably well positioned on utility maps, but underground connections (pipes, conduits, cables, etc.) are usually shown as straight lines connecting the surface features. During original construction of such utilities, trenching may have been designed as a series of straight lines, but, in actuality, buried obstructions such as boulders, unstable soil or unmapped existing utilities necessitated deviation from the designed trench alignment. In many instances, as-built surveys were never done after construction, and the design map, without any notation of as-constructed alignment changes, became the only map recording the location of the constructed utilities.

In well-developed areas, it may not be realistic to attempt to locate all utilities during the design phase of a project without a prohibitive amount of investigation, which is beyond the time and cost limitations of the designer's budget. However, the designer must perform a diligent effort to minimise surprises during excavation and construction. Again, the level of due diligence depends on the method of excavation (cut-and-cover, or mined tunnel), the depth of the tunnel, and the number, size and location of proposed shafts.

Identification of Underground Structures and Other Obstacles

Often, particularly in dense urban areas, other underground structures may exist that may impact the alignment and profile of the proposed road tunnel, and will dictate the need for structure protection measures during construction. These existing underground structures may include transit and railroad tunnels, other road tunnels, underground pedestrian passageways, building vaults, existing or abandoned marine structures (bulkheads, piers, etc.), and existing or abandoned structure foundations. Other underground obstructions may include abandoned temporary shoring systems, soil treatment areas, and soil or rock anchors that were used for temporary or permanent support of earth retaining structures. Initial surveys for the project should therefore include a survey of existing and past structures using documents from city and state agencies, and building owners. In addition, historical maps and records should be reviewed to assess the potential for buried abandoned structures.

Structure Preconstruction Survey

Structures located within the zone of potential influence may experience a certain amount of vertical and lateral movement as a result of soil movement caused by tunnel excavation and construction in close proximity (e.g. cut-and-cover excavation, shallow soft ground tunnelling, etc.). If the anticipated movement may induce potential damage to a structure, some protection measures will be required, and a detailed preconstruction survey of the structure should be performed. Preconstruction survey should ascertain all pertinent facts of pre-existing conditions, and identify features and locations for further monitoring.

Geologic Mapping

After collecting and reviewing existing geologic maps, aerial photos, references, and the results of a preliminary site reconnaissance, surface geologic mapping of available rock outcrops should be performed by an experienced engineering geologist to obtain detailed, site-specific information on rock quality and structure. Geologic mapping collects local, detailed geologic data systematically, and is used to characterise and document the condition of rock mass or outcrop for rock mass classification such as:

- Discontinuity type
- Discontinuity orientation
- Discontinuity infilling
- Discontinuity spacing
- Discontinuity persistence
- Weathering

The International Society of Rock Mechanics (ISRM) has suggested quantitative measures for describing discontinuities (ISRM 1981). It provides standard descriptions for factors such as persistence, roughness, wall strength, aperture, filling, seepage, and block size. Where necessary, it gives suggested methods for measuring these parameters so that the discontinuity can be characterised in a constant manner that allows comparison.

By interpreting and extrapolating all these data, the geologist should have a better understanding of the rock conditions likely to be present along the proposed tunnel and at the proposed portal and shaft excavations. The collected mapping data can be used in stereographic projections for statistical analysis using appropriate computer software (e.g., DIPS), in addition to the data obtained from the subsurface investigations.

In addition, the following surface features should also be observed and documented during the geologic mapping program:

- Slides, new or old, particularly in proposed portal and shaft areas
- Faults
- Rock weathering
- Sinkholes and karstic terrain
- Groundwater springs
- Volcanic activity
- Anhydrite, gypsum, pyrite, or swelling shales
- Stress relief cracks
- Presence of talus or boulders
- Thermal water (heat) and gas

The mapping data will also help in targeting subsurface investigation borings and in situ testing in areas of observed variability and anomalies.

Subsurface Investigations

Ground conditions including geological, geotechnical, and hydrological conditions, have a major impact on the planning, design, construction and cost of a road tunnel, and often determine its feasibility and final route. Fundamentally, subsurface investigation is the most important type of investigations to obtain ground conditions, as it is the principal means for:

- Defining the subsurface profile (i.e. stratigraphy, structure, and principal soil and rock types)
- Determining soil and rock material properties and mass characteristics;

- Identify geological anomalies, fault zones and other hazards (squeezing soils, methane gas, etc.)
- Defining hydrogeological conditions (groundwater levels, aquifers, hydrostatic pressures, etc.); and
- Identifying potential construction risks (boulders, etc.).

Subsurface investigations typically consist of borings, sampling, in situ testing, geophysical investigations, and laboratory material testing. The principal purposes of these investigation techniques are summarised below:

- Borings are used to identify the subsurface stratigraphy, and to obtain disturbed and undisturbed samples for visual classification and laboratory testing;
- In situ tests are commonly used to obtain useful engineering and index properties by testing the material in place to avoid the disturbance inevitably caused by sampling, transportation and handling of samples retrieved from boreholes; in situ tests can also aid in defining stratigraphy;
- Geophysical tests quickly and economically obtain subsurface information (stratigraphy and general engineering characteristics) over a large area to help define stratigraphy and to identify appropriate locations for performing borings; and
- Laboratory testing provides a wide variety of engineering properties and index properties from representative soil samples and rock core retrieved from the borings.

Unlike other highway structures, the ground surrounding a tunnel can act as a supporting mechanism, or loading mechanism, or both, depending on the nature of the ground, the tunnel size, and the method and sequence of constructing the tunnel. Thus, for tunnel designers and contractors, the rock or soil surrounding a tunnel is a construction material, just as important as the concrete and steel used on the job. Therefore, although the above subsurface investigative techniques are similar (or identical) to the ones used for foundation design as specified in AASHTO 2006 Interim and in accordance with appropriate ASTM or AASHTO standards, the geological and geotechnical focuses for underground designs and constructions can be vastly different. In addition to typical geotechnical, geological, and geo-hydrological data, subsurface investigation for a tunnel project must consider the unique needs for different tunnelling methods, i.e.

cut-and-cover, drill-and-blast, bored, sequential excavation, and immersed. Table 3-2 shows other special considerations for various tunnelling methods.

Subsurface investigations must be performed in phases to better economize the program.

Nonetheless, they are primarily performed during the design stage of the project, with much of the work typically concentrated in the preliminary design phase of a project.

These investigations provide factual information about the distribution and engineering characteristics of soil, rock and groundwater at a site, allowing an understanding of the existing conditions sufficient for developing an economical design, determining a reliable construction cost estimate, and reducing the risks of construction.

The specific scope and extent of the investigation must be appropriate for the size of the project and the complexity of the existing geologic conditions; must consider budgetary constraints; and must be consistent with the level of risk considered acceptable to the client. To ensure the collected data can be analysed correctly throughout the project, the project coordinate system and vertical datum should be established early on and the boring and testing locations must be surveyed, at least by hand-held GPS equipment. Photographs of the locations should be maintained as well.

Since unanticipated ground conditions are most often the reason for costly delays, claims and disputes during tunnel construction, a project with a more thorough subsurface investigation program would likely have fewer problems and lower final cost.

Therefore, ideally, the extent of an exploration program should be based on specific project requirements and complexity, rather than strict budget limits. However, for most road tunnels, especially tunnels in mountainous areas or for water crossings, the cost for a comprehensive subsurface investigation may be prohibitive. The challenge to geotechnical professionals is to develop an adequate and diligent subsurface investigation program that can improve the predictability of ground conditions within a reasonable budget and acceptable level of risk. It is important that the involved parties have a common understanding of the limitations of geotechnical investigations, and be aware of the inevitable risk of not being able to completely define existing geological conditions.

Table 1: Special Investigation Needs Related to Tunnelling Methods

Cut and Cover	Plan exploration to obtain design parameters for excavation support, and specifically define conditions closely enough to reliably determine best and most cost-effective location to change from cut-and-cover to true tunnel mining construction.
Drill and Blast	Data needed to predict stand-up time for the size and orientation of tunnel.
Rock Tunnel Boring Machine	Data required to determine cutter costs and penetration rate is essential. Need data to predict stand-up time to determine if open-type machine will be ok or if full shield is necessary. Also, water inflow is very important.
Roadheader	Need data on jointing to evaluate if roadheader will be plucking out small joint blocks or must grind rock away. Data on hardness of rock is essential to predict cutter/pick costs.
Shielded Soft Ground Tunnel Boring Machine	Stand-up time is important to face stability and the need for breasting at the face as well as to determine the requirements for filling tail void. Need to fully characterise all potential mixed-face conditions.
Pressurized-Face Tunnel Boring Machine	Need reliable estimate of groundwater pressures and of strength and permeability of soil to be tunnelled. Essential to predict size, distribution and amount of boulders. Mixed-face conditions must be fully characterised.
Compressed Air	Borings must not be drilled right on the alignment and must be well grouted so that compressed air will not be lost up old bore hole in case tunnel encounters old boring
Solution-Mining	Need chemistry to estimate rate of leaching and undisturbed core in order to conduct long-term creep tests for cavern stability analyses.
Sequential Excavation Method/NATM	Generally requires more comprehensive geotechnical data and analysis to predict

	behaviour and to classify the ground conditions and ground support systems into four or five categories based on the behaviour.
Immersed Tube	Need soil data to reliably design dredged slopes and to predict rebound of the dredged trench and settlement of the completed immersed tube structure. Testing should emphasize rebound modulus (elastic and consolidation) and unloading strength parameters. Usual softness of soil challenges determination of strength of soil for slope and bearing evaluations. Also need exploration to assure that all potential obstructions and/or rock ledges are identified, characterised, and located. Any contaminated ground should be fully characterised.
Jacked Box Tunnelling	Need data to predict soil skin friction and to determine the method of excavation and support needed at the heading
Portal Construction	Need reliable data to determine most cost-effective location of portal and to design temporary and final portal structure. Portals are usually in weathered rock/soil and sometimes in strain-relief zone.
Construction Shafts	Should be at least one boring at every proposed shaft location.
Access, Ventilation, or Other Permanent Shafts	Need data to design the permanent support and groundwater control measures. Each shaft deserves at least one boring.

Table 2: Geotechnical Investigation

Hard or Abrasive Rock	• Difficult and expensive for TBM or roadheader. Investigate, obtain samples, and conduct lab tests to provide parameters needed to predict rate of advance and cutter costs.
Mixed Face	• Especially difficult for wheel type TBM
	• Particularly difficult tunnelling condition in soil and in rock. Should be characterised carefully to determine nature and behaviour of mixed-face and approximately length of tunnel likely to be affected for each mixed-face condition.

Karst	• Potentially large cavities along joints, especially at intersection of master joint systems; small but sometimes very large and very long caves capable of undesirably large inflows of groundwater.
Gypsum	• Potential for soluble gypsum to be missing or to be removed because of change of groundwater conditions during and after construction.
Salt or Potash	• Creep characteristics and, in some cases, thermal-mechanical characteristics are very important
Saprolite	• Investigate for relict structure that might affect behaviour
	• Depth and degree of weathering; important to characterise especially if tunnelling near rock-soil boundary
	• Different rock types exhibit vastly differing weathering profiles
High In-Situ Stress	• Could strongly affect stand-up time and deformation patterns both in soil and rock tunnels. Should evaluate for rock bursts or popping rock in particularly deep tunnels
Low In-Situ Stress	• Investigate for open joints that dramatically reduce rock mass strength and modulus and increase permeability. Often potential problem for portals in downcut valleys and particularly in topographic "noses" where considerable relief of strain could occur.
	• Conduct hydraulic jacking and hydrofracture tests.
Hard Fissured or Slickensided Soil	• Lab tests often overestimate mass physical strength of soil. Large-scale testing and/or exploratory shafts/adits may be appropriate
Gassy Ground-always test	• Methane (common)
for hazardous gases	• H2S
Adverse Geological Features	• Faults
	• Known or suspected active faults. Investigate to determine location and estimate likely ground motion

	• Inactive faults but still sources of difficult tunnelling condition
	• Faults sometimes act as dams and other times as drainage paths to groundwater
	• Fault gouge sometimes a problem for strength and modulus
	• High temperature groundwater
	• Always collect samples for chemistry tests
	• Sedimentary Formations
	• Frequently highly jointed
	• Concretions could be problem for TBM
Adverse Geological Features (Continued)	• Groundwater
	• Groundwater is one of the most difficult and costly problems to control. Must investigate to predict groundwater as reliably as possible
	• Site characterisation should investigate for signs of and nature of:
	• Groundwater pressure
	• Groundwater flow
	• Artesian pressure
	• Multiple aquifers
	• Higher pressure in deeper aquifer
	• Groundwater perched on top of impermeable layer in mixed face condition
	• Ananalous or abrupt
	• Aggressive groundwater
	• Soluble sulfates that attack concrete and shotcrete
	• Pyrites
	• Acidic
	• Lava or Volcanic Formation
	• Flow tops and flow bottoms frequently are very permeable and difficult tunnelling ground
	• Lava Tubes
	• Vertical borings do not disclose the nature of columnar jointing. Need inclined borings

	• Potential for significant groundwater flows from columnar jointing • Boulders (sometimes nests of boulders) frequently rest at base of strata • Cobbles and boulders not always encountered in borings which could be misleading. • Should predict size, number, and distribution of boulders on basis of outcrops and geology • Beach and Fine Sugar Sands • Very little cohesion. Need to evaluate stand-up time. • Glacial deposits • Boulders frequently associated with glacial deposits. Must actively investigate for size, number, and distribution of boulders. • Some glacial deposits are so hard and brittle that they are jointed and ground behaviour is affected by the joining as well as properties of the matrix of the deposit • Permafrost and frozen soils • Special soil sampling techniques required • Thermal-mechanical properties required
Manmade Features	
	• Contaminated groundwater/soil • Check for movement of contaminated plume caused by changes in groundwater regime as a result of construction • Existing Obstructions • Piles • Previously constructed tunnels • Tiebacks extending out into sheet • Existing Utilities • Age and condition of overlying or adjacent utilities within zone of influence

A general approach to control the cost of subsurface investigations while obtaining the information necessary for design and construction would include a) phasing the investigation, to better match and limit the scope of the investigation to the specific needs for each phase of the project, and b) utilising existing information and the results of geologic mapping and geophysical testing to more effectively select locations for investigation. Emphasis can be placed first on defining the local geology, and then on increasingly greater detailed characterisation of the subsurface conditions and predicted ground behaviour. Also, subsurface investigation programs need to be flexible and should include an appropriate level of contingency funds to further assess unexpected conditions and issues that may be exposed during the planned program. Failure to resolve these issues early may lead to costly redesign or delays, claims and disputes during construction.

Unless site constraints dictate a particular alignment, such as within a confined urban setting, few projects are constructed precisely along the alignment established at the time the initial boring program is laid out. This should be taken into account when developing and budgeting for geotechnical investigations, and further illustrates the need for a phased subsurface investigation program.

Test Borings and Sampling

Vertical and Inclined Test Borings: Vertical and slightly inclined test borings and soil/rock sampling are key elements of any subsurface investigations for underground projects. The location, depth, sample types and sampling intervals for each test boring must be selected to match specific project requirements, topographic setting and anticipated geological conditions. Various field testing techniques can be performed in conjunction with the test borings as well. Refer to FHWA Reference Manual for Subsurface Investigations (FHWA, 2002b) and GEC 5 (FHWA, 2002a) for guidance regarding the planning and conduct of subsurface exploration programs.

This table presents general guidelines from AASHTO (1988) for determining the spacing of boreholes for tunnel projects:

Table 3: Guidelines for Vertical/Inclined Borehole Spacing

Ground Conditions	*Typical Borehole Spacing (feet)*
Cut-and-Cover Tunnels	100 to 300
Rock Tunnelling	
Adverse Conditions	50 to 200
Favourable Conditions	500 to 1000
Soft Ground Tunnelling	
Adverse Conditions	50 to 100

Favourable Conditions	300 to 500
Mixed Face Tunnelling	
Adverse Conditions	25 to 50
Favourable Conditions	50 to 75

The above guideline can be used as a starting point for determining the number and locations of borings. However, especially for a long tunnel through a mountainous area, under a deep water body, or within a populated urban area, it may not be economically feasible or the time sufficient to perform borings accordingly. Therefore, engineering judgment will need to be applied by a licensed and experienced geotechnical professional to adapt the investigation program.

In general, borings should be extended to at least *1.5 tunnel diametres* below the proposed tunnel invert. However, if there is uncertainty regarding the final profile of the tunnel, the borings should extend at least two or three times the tunnel diametre below the preliminary tunnel invert level. Borings at shafts should extend at least *1.5 times the depth of the shaft* for design of the shoring system and shaft foundation, especially in soft soils.

Horizontal and Directional Boring/Coring

Horizontal boreholes along tunnel alignments provide a continuous record of ground conditions and information which is directly relevant to the tunnel alignment. Although the horizontal drilling and coring cost per linear feet may be much higher than the conventional vertical/ inclined borings, a horizontal borings can be more economical, especially for investigating a deep mountainous alignment, since one horizontal boring can replace many deep vertical conventional boreholes and avoid unnecessary drilling of overburden materials and disruption to the ground surface activities, local community and industries.

A deep horizontal boring will need some distance of inclined drilling through the overburden and upper materials to reach to the depth of the tunnel alignment. Typically the inclined section is stabilised using drilling fluid and casing and no samples are obtained. Once the bore hole reached a horizontal alignment, coring can be obtained using HQ triple tube core barrels.

Sampling - Overburden Soil

Standard split spoon (disturbed) soil samples (ASTM D-1586) are typically obtained at intervals not greater than 5 feet and at changes in strata. Continuous sampling from one diametre above the tunnel crown to one diametre below the tunnel invert is advised to better define the stratification and materials within this zone if within soil or intermediate geomaterial.

In addition, undisturbed tube samples should be obtained in each cohesive soil stratum encountered in the borings; where a thick stratum of cohesive soil is present, undisturbed samples should be obtained at intervals not exceeding 15 ft. Large diametre borings or rotosonic type borings can be considered to obtain special samples for classification and testing.

Figure: Rotosonic Sampling for a CSO Tunnel Project at Portland, Oregon.

Sampling – Rock Core

In rock, continuous rock core should be obtained below the surface of rock, with a minimum NX-size core (diametre of 2.16 inch or 54.7 mm). Double and triple tube core barrels should be used to obtain higher quality core more representative of the in situ rock. For deeper holes, coring should be performed with the use of wire-line drilling equipment to further reduce potential degradation of the recovered core samples. Core runs should be limited to a maximum length of 10 ft in moderate to good quality rock, and 5 ft in poor quality rock.

The rock should be logged soon after it was extracted from the core barrel. Definitions and terminologies used in logging rock cores are presented in Appendix B. Primarily, the following information is recommended to be noted for each core run on the rock coring logs:

- Depth of core run
- Core recovery in inches and percent
- Rock Quality Designation (RQD) percent
- Rock type, including colour texture, degree of weathering and hardness

- Character of discontinuities, joint spacing, orientation, roughness and alteration
- Nature of joint infilling materials

In addition, drilling parameters, such as type of drilling equipment, core barrel and casing size, drilling rate, and groundwater level logged in the field can be useful in the future. Typical rock coring logs for tunnel design purpose are included in Appendix B.

Borehole Sealing

All borings should be properly sealed at the completion of the field exploration, if not intended to be used as monitoring wells. This is typically required for safety considerations and to prevent cross contamination of soil strata and groundwater. However, boring sealing is particularly important for tunnel projects since an open borehole exposed during tunnelling may lead to uncontrolled inflow of water or escape of slurry from a slurry shield TBM or air from a compressed air tunnel.

In many parts of the country, methods used for sealing of boreholes are regulated by state agencies. FHWA-NHI-035 "Workbook for Subsurface Investigation Inspection Qualification" (FHWA, 2006a) offers general guidelines for borehole sealing. National Cooperative Highway Research Program Report No. 378 (Lutenegger et al., 1995), titled "Recommended Guidelines for Sealing Geotechnical Holes," contains extensive information on sealing and grouting boreholes.

Backfilling of boreholes is generally accomplished using a grout mixture by pumping the grout mix through drill rods or other pipes inserted into the borehole. In boreholes where groundwater or drilling fluid is present, grout should be tremied from the bottom of the borehole. Provision should be made to collect and dispose of all drill fluid and waste grout. Holes in pavement and slabs should be patched with concrete or asphaltic concrete, as appropriate.

Test Pits

Test pits are often used to investigate the shallow presence, location and depth of existing utilities, structure foundations, top of bedrock and other underground features that may interfere or be impacted by the construction of shafts, portals and cut-and-cover tunnels. The depth and size of test pits will be dictated by the depth and extent of the feature being exposed. Except for very shallow excavations, test pits will typically require sheeting and shoring to provide positive ground support and ensure the safety of individuals entering the excavation in compliance with OSHA and other regulatory requirements. The conditions exposed in test pits, including the existing soil and rock materials, groundwater observations, and utility and structure elements are documented by written records and photographs, and representative materials are sampled for future visual examination and laboratory testing. The excavation pits are then generally backfilled with excavation spoil, and the backfill is compacted to avoid excessive future settlement. Tampers and rollers may be used to facilitate compaction of the backfill. The ground surface or pavement is then typically restored using materials and thickness dimension matching the adjoining areas.

Soil and Rock Identification and Classification

Soil Identification and Classification: It is important to distinguish between visual identification and classification to minimise conflicts between general visual identification of soil samples in the field versus a more precise laboratory evaluation supported by index tests. Visual descriptions in the field are often subjected to outdoor elements, which may influence results. It is important to send the soil samples to a laboratory for accurate visual identification by a geologist or technician experienced in soils work, as this single operation will provide the basis for later testing and soil profile development.

During progression of a boring, the field personnel should describe the sample encountered in accordance with the ASTM D 2488, Practice for Description and Identification of Soils (Visual-Manual Procedure), which is the modified Unified Soil Classification System (USCS). For detailed field identification procedures for soil samples readers are referred to FHWA-NHI-035 "Workbook for Subsurface Investigation Inspection Qualification".

For the most part, field classification of soil for a tunnel project is similar to that for other geotechnical applications except that special attention must be given to accurately defining and documenting soil grain size characteristics and stratification features since these properties may have greater influence on the ground and groundwater behaviour during tunnelling than they may have on other types of construction, such as for foundations, embankments and cuts. Items of particular importance to tunnel projects are listed below:

- Groundwater levels (general and perched levels), evidence of ground permeability (loss of drilling fluid; rise or drop in borehole water level; etc.), and evidence of artesian conditions
- Consistency and strength of cohesive soils

- Composition, gradation and density of cohesionless soils
- Presence of lenses and layers of higher permeability soils
- Presence of gravel, cobbles and boulders, and potential for nested boulders
- Maximum cobble/boulder size from coring and/or large diametre borings (and also based on understanding of local geology), and the unconfined compressive strength of cobbles/boulders (from field index tests and laboratory testing of recovered samples)
- Presence of cemented soils
- Presence of contaminated soil or groundwater

All of the above issues will greatly influence ground behaviour and groundwater inflow during construction, and the selection of the tunnelling equipment and methods.

Rock Identification and Classification

In rock, rock mass characteristics and discontinuities typically have a much greater influence on ground behaviour during tunnelling and on tunnel loading than the intact rock properties. Therefore, rock classification needs to be focused on rock mass characteristics, as well as its origin and intact properties for typical highway foundation application. Special intact properties are important for tunnelling application particularly for selecting rock cutters for tunnel boring machines and other types of rock excavators, and to predict cutter wear.

Typical items included in describing general rock lithology include:

- General rock type
- Colour
- Grain size and shape
- Texture (stratification, foliation, etc.)
- Mineral composition
- Hardness
- Abrasivity
- Strength
- Weathering and alteration

Rock discontinuity descriptions typically noted in rock classification include:

- Predominant joint sets (with strike and dip orientations)
- Joint roughness
- Joint persistence
- Joint spacing
- Joint weathering and infilling

Other information typically noted during subsurface rock investigations include:

- Presence of faults or shear zones
- Presence of intrusive material (volcanic dikes and sills)
- Presence of voids (solution cavities, lava tubes, etc.)
- Groundwater levels, and evidence of rock mass permeability (loss of drilling fluid; rise or drop in borehole water level; etc.)

Index properties obtained from inspection of the recovered rock core include core recovery (i.e., the recovered core length expressed as a percentage of the total core run length), and Rock Quality Designation or RQD (the combined length of all sound and intact core segments equal to or greater than 4 inches in length, expressed as a percentage of the total core run length).

Often, materials encountered during subsurface investigations represent a transitional (intermediate) material formed by the in place weathering of rock. Such conditions may sometimes present a complex condition with no clear boundaries between the different materials encountered. Tunnelling through the intermediate geomaterial (IGM), in some cases referred as mixed-face condition, can be extremely difficult especially when groundwater is present. In the areas where tunnel alignment must cross this transition zone, the subsurface investigation is conducted much as for rock, and when possible cores are retrieved and classified, and representative intact pieces of rock should be tested.

Field Testing Techniques (Pre-Construction)

Field testing for subsurface investigations includes two general categories of tests:

- a) In situ tests
- b) Geophysical testing

In situ tests are used to directly obtain field measurements of useful soil and rock engineering properties. Geophysical tests, the second general category of field tests, are indirect methods of exploration in which changes in certain physical characteristics such as magnetism, density, electrical resistivity, elasticity, or a combination of these are used as an aid in developing subsurface information. There are times that two testing methods can be performed from a same apparatus, such as using seismic CPT.

In situ Testing

In situ tests are used to directly obtain field measurements of useful soil and rock engineering properties. In soil, in situ testing include both index type tests, such as the Standard Penetration Test (SPT) and tests that determine the physical properties of the ground, such as shear strength from cone penetration Tests (CPT) and ground deformation properties from pressure metre tests (PMT). In situ test methods in soil commonly used in the U.S. and their applications and limitations are summarised.

One significant property of interest in rock is its in situ stress condition. Horizontal stresses of geological origin are often locked within the rock masses, resulted in a stress ratio (K) often higher than the number predicted by elastic theory. Depending on the size and orientation of the tunnelling, high horizontal stresses may produce favourable compression in support and confinement, or induce popping or failure during and after excavation. Principally, two different general methods are common to be employed to measure the in situ stress condition: hydraulic fracturing and overcoring. Note that in situ stress can only be measured accurately within a fair or better rock condition. However, since weak rocks are unable to support large deviatoric stress differences, the lateral and vertical stresses tend to equalize over geologic time.

Laboratory Testing

Soil Testing Detailed soil laboratory testing is required to obtain accurate information including classification, characteristics, stiffness, strength, etc. for design and modelling purposes. Testing are performed on selected representative samples in accordance with ASTM standards. Rock Testing Standard rock testing evaluate physical properties of the rock included density and mineralogy (thin-section analysis). The mechanical properties of the intact rock core included uniaxial compressive strength, tensile strength, static and dynamic elastic constants, hardness, and abrasitivity indices.

In addition, specialised tests for assessing TBM performance rates are required including three drillability and boreability testing, namely, Drilling Rate Index (DRI), Bit Wear Index (BWI), and Cutter Life Index (CLI). Table 3-9 summarises common rock laboratory testing for tunnel design purposes.

It is desirable to preserve the rock cores retrieved from the field properly for years until the construction is completed and disputes/ claims are settled. Common practice is to photograph the rock cores in core boxes and possibly scan the core samples for review by designers and contractors.

Parameter	***Test Method***
Index properties	• Density
	• Porosity
	• Moisture Content
	• Slake Durability
	• Swelling Index
	• Point Load Index
	• Hardness
	• Abrasivity
Strength	• Uniaxial compressive strength
	• Triaxial compressive strength
	• Tensile strength (Brazilian) Shear strength of joints
Deformability	• Young's modulus Poisson's ratio
Time dependence	• Creep characteristics
Permeability	• Coefficient of permeability
Mineralogy and grain sizes	• Thin-sections analysis
	• Differential thermal analysis
	• X-ray diffraction

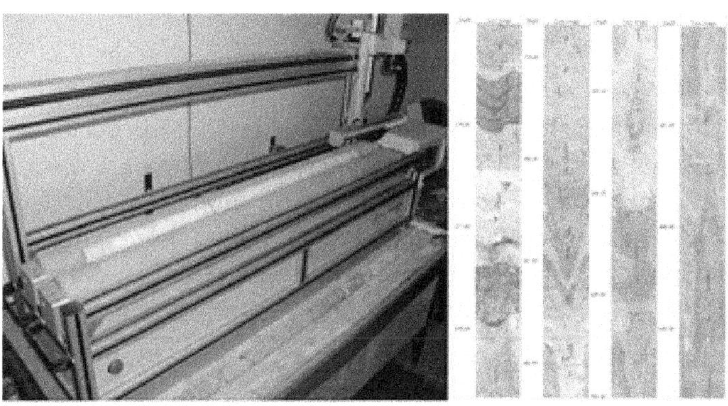

Groundwater Investigation

Groundwater is a major factor for all types of projects, but for tunnels groundwater is a particularly critical issue since it may not only represent a large percentage of the loading on the final tunnel lining, but also it largely determines ground behaviour and stability for soft ground tunnels; the inflow into rock tunnels; the method and equipment selected for tunnel construction; and the long-term performance of the completed structure. Accordingly, for tunnel projects, special attention must be given to defining the groundwater regime, aquifers, and sources of water, any perched or artesian conditions, water quality and temperature, depth to groundwater, and the permeability of the various materials that may be encountered during tunnelling. Related considerations include the potential impact of groundwater lowering on settlement of overlying and nearby structures, utilities and other facilities; other influences of dewatering on existing structures (e.g., accelerated deterioration of exposed timber piles); pumping volumes during construction; decontamination/treatment measures for water discharged from pumping; migration of existing soil and groundwater contaminants due to dewatering; the potential impact on water supply aquifers; and seepage into the completed tunnel; to note just a few:

Groundwater investigations typically include most or all of the following elements:

- Observation of groundwater levels in boreholes
- Assessment of soil moisture changes in the boreholes
- Groundwater sampling for environmental testing
- Installation of groundwater observation wells and piezometres
- Borehole permeability tests (rising, falling and constant head tests; packer tests, etc.)
- Geophysical testing
- Pumping tests

During subsurface investigation drilling and coring, it is particularly important for the inspector to note and document any groundwater related observations made during drilling or during interruptions to the work when the borehole has been left undisturbed. Even seemingly minor observations may have an important influence on tunnel design and ground behaviour during construction.

Groundwater observation wells are used to more accurately determine and monitor the static water table. Since observation wells are generally not isolated within an individual zone or stratum they provide only a general indication of the groundwater table, and are therefore more suitable for sites with generally uniform subsurface conditions. In stratified soils with two or more aquifers, water pressures may vary considerably with depth. For such variable conditions, it is generally more appropriate to use piezometres. Piezometres have seals that isolate the screens or sensors within a specific zone or layer within the soil profile, providing a measurement of the water pressure within that zone.

Observation wells and piezometres should be monitored periodically over a prolonged period of time to provide information on seasonal variations in groundwater levels. Monitoring during construction provides important information on the influence of tunnelling on groundwater levels, forming an essential component of construction control and any protection program for existing structures and facilities. Local and state jurisdictions may impose specific requirements for permanent observation wells and piezometres, for documenting both temporary and permanent installations, and for closure of these installations.

Borehole Permeability Testing

Borehole permeability tests provide a low cost means for assessing the permeability of soil and rock. The principal types of tests include falling head, rising head and constant head tests in soil, and packer tests in rock, as described below. Additional information regarding the details and procedures used for performing and interpreting these borehole permeability tests are presented by FHWA (2002b). Borehole tests are particularly beneficial in sands and gravels since samples of such materials would be too disturbed to use for laboratory permeability tests. A major limitation of these tests, however, is that they assess soil conditions only in the immediate vicinity of the borehole, and the results do not reflect the influence of water recharge sources or soil stratification over a larger area.

Borehole permeability tests are performed intermittently as the borehole is advanced. Holes in which permeability tests will be performed should be drilled with water to avoid the formation of a filter cake on the sides of the borehole from drilling slurry. Also, prior to performing the permeability test the hole should be flushed with clear water until all sediments are removed from the hole (but not so much as would be done to establish a water well).

In soil, either rising head or falling head tests would be appropriate if the permeability is low enough to permit accurate determination of water level versus time. In the falling head test, where the flow is from the hole to the surrounding soil, there is risk of clogging of the soil pores by sediments in the test water. In the rising head tests, where water flows from the surrounding soil into the hole, there is a risk of the soil along the test length becoming loosened or quick if the seepage gradient is too large. If a rising head test is used, the hole should be sounded at the end of the test to determine if the hole has collapsed or heaved. Generally, the rising head test is the preferred test method. However, in cases where the permeability is so high as to preclude accurate measurement of the rising or falling water level, the constant head test should be used.

Pressure, or "Packer," tests are performed in rock by forcing water under pressure into the rock surrounding the borehole. Packer tests determine the apparent permeability of the rock mass, and also provide a qualitative assessment of rock quality. These tests can also be used before and after grouting to assess the effectiveness of grouting on rock permeability and the strength of the rock mass. The test is performed by selecting a length of borehole for testing, then inflating a cylindrical rubber sleeve ("packer") at the top of the test zone to isolate the section of borehole being tested. Packer testing can thus be performed intermittently as the borehole is advanced. Alternatively, testing can be performed at multiple levels in a completed borehole by using a double packer system in which packers are positioned and inflated at both the top and bottom of the zone being tested, as illustrated. Once the packer is inflated to seal off the test section, water is pumped under pressure to the test zone, while the time and volume of water pumped at different pressures are recorded.

Pumping Tests

Continuous pumping tests are used to determine the water yield of individual wells and the permeability of subsurface materials in situ over an extended area. These data provide useful information for predicting inflows during tunnelling; the quantity of water that may need to be pumped to lower groundwater levels; and the radius of influence for pumping operations; among others. The test consists of pumping water from a well or borehole and observing the effect on the water table with distance and time by measuring the water levels in the hole being pumped as well as in an array of observation wells at various distances around the pumping well. The depth of the test well will depend on the depth and thickness of the strata being investigated, and the number, location and depth of the observation wells or piezometres will depend on the anticipated shape of the groundwater surface after drawdown.

Environmental Issues

Although tunnels are generally considered environmentally-friendly structures, certain short-term environmental impacts during construction are unavoidable. Long-term impacts from the tunnel itself, and from portals, vent shafts and approaches on local communities, historic sites, wetlands, and other aesthetically, environmentally, and ecologically sensitive areas must be identified and investigated thoroughly during the project planning and feasibility stages, and appropriately addressed in environmental studies and design.

Early investigation and resolution of environmental issues is an essential objective for any underground project since unanticipated conditions discovered later during design or construction could potentially jeopardize the project.

The specific environmental data needed for a particular underground project very much depend on the geologic and geographic environment and the functional requirement of the underground facility. Some common issues can be stated, however, and are identified below in the form of a checklist:

- Existing infrastructure, and obstacles underground and above
- Surface structures within area of influence
- Land ownership and uses (public and private)
- Ecosystem habitat impacts
- Contaminated ground or groundwater
- Long-term impacts to groundwater levels, aquifers and water quality
- Control of runoff and erosion during construction
- Naturally gassy ground, or groundwater with deleterious chemistry
- Access constraints for potential work sites and transport routes
- Sites for muck transport and disposal
- Noise and vibrations from construction operations, and from future traffic at approaches to the completed tunnel

- Air quality during construction, and at portals, vent shafts and approaches of the completed tunnel
- Maintenance of vehicular traffic and transit lines during construction
- Maintenance of utilities and other existing facilities during construction
- Access to residential and commercial properties
- Pest control during construction
- Long-term community impacts
- Long-term traffic impacts
- Temporary and permanent easements
- Tunnel fire life safety and security
- Legal and environmental constraints, enumerated in environmental statements or reports, or elsewhere

Seismicity

The release of energy from earthquakes sends seismic acceleration waves travelling through the ground. Such transient dynamic loading instantaneously increases the shear stresses in the ground and decreases the volume of voids within the material which leads to an increase in the pressure of fluids (water) in pores and fractures. Thus, shear forces increase and the frictional forces that resist them decrease. Other factors also can affect the response of the ground during earthquakes.

- Distance of the seismic source from the project site.
- Magnitude of the seismic accelerations.
- Earthquake duration.
- Subsurface profile.
- Dynamic characteristics and strengths of the materials affected.

In addition to the distance of the seismic source to the project site, and the design (anticipated) time history, duration and magnitude of the bedrock earthquake, the subsurface soil profile can have a profound effect on earthquake ground motions including the intensity, frequency content, and duration of earthquake shaking. Amplification of peak bedrock acceleration by a factor of four or more has been attributed to the response of the local soil profile to the bedrock ground motions.

The ground accelerations associated with seismic events can induce significant inertial forces that may lead to instability and permanent deformations (both vertically and laterally) of tunnels and portal slopes. In addition, during strong earthquake shaking, saturated cohesionless soils may experience a sudden loss of strength and stiffness, sometime resulting in loss of bearing capacity, large permanent lateral displacements, landslides, and/or seismic settlement of the ground. Liquefaction beneath and in the vicinity of a portal slope can have severe consequences since global instability in forms of excessive lateral displacement or lateral spreading failure may occur as a result. Readers are referred to FHWA publication "Geotechnical Earthquake Engineering" by Kavazanjian, et al. (1998) for a detailed discussion of this topic.

Additional Investigations During Construction

For tunnelling projects it is generally essential to perform additional subsurface investigations and ground characterisation during construction. Such construction phase investigations serve a number of important functions, providing information for:

- Contractor design and installation of temporary works
- Further defining anomalies and unanticipated conditions identified after the start of construction
- Documenting existing ground conditions for comparison with established baseline conditions, thereby forming the basis of any cost adjustments due to differing site conditions
- Assessing ground and groundwater conditions in advance of the tunnel heading to reduce risks and improve the efficiency of tunnelling operations
- Determining the initial support system to be installed, and the locations where the support system can be changed
- Assessing the response of the ground and existing structures and utilities to tunnelling operations
- Assessing the groundwater table response to dewatering and tunnelling operations
- Determining the location and depth of existing utilities and other underground facilities

A typical construction phase investigation program would likely include some or all of the following elements:

- Subsurface investigation borings and probings from the ground surface
- Test pits
- Additional groundwater observation wells and/or piezometres
- Additional laboratory testing of soil and rock samples
- Geologic mapping of the exposed tunnel face
- Geotechnical instrumentation
- Probing in advance of the tunnel heading from the face of the tunnel
- Pilot Tunnels
- Environmental testing of soil and groundwater samples suspected to be contaminated or otherwise harmful

Some of the above investigation elements, such as geotechnical instrumentation, may be identified as requirements of the contract documents, while others, such as additional exploratory borings, may be left to the discretion of the contractor for their benefit and convenience. Tunnel face mapping and groundwater monitoring should be required elements for any tunnel project since the information obtained from these records will form the basis for evaluating the merits of potential differing site condition claims.

Geologic Face Mapping

With open-face tunnelling methods, including the sequential excavation method (SEM), open-face tunnelling shield in soil, and the drill-and-blast method in rock, all or a large portion of the tunnel face will be exposed, allowing a visual assessment of the existing ground and groundwater conditions. In such cases, the exposed face conditions are documented in cross-section sketches (face mapping) drawn at frequent intervals as the tunnel advances. Information typically included in these face maps include the station location for the cross-section; the date and time the face mapping was prepared; the name of the individual who prepared the face map; classification of each type of material observed; the location of interface boundaries between these materials; rock jointing including orientation of principal joints and joint descriptions; shear zones; observed seepage conditions and their approximate locations on the face; observed ground behaviour noting particularly the location of any instability or squeezing material at the face; the location of any boulders, piling or other obstructions; the location of any grouted or cemented material; and any other significant observations. In rock tunnels where the perimetre rock is left exposed, sketches presenting similar information can be prepared for the tunnel walls and roof. All mapping should be prepared by a geologist or geological engineer knowledgeable of tunnelling and with soil and rock classification.

The face maps can be used to accurately document conditions exposed during tunnelling, and to develop a detailed profile of subsurface conditions along the tunnel horizon. However, there are limitations and considerable uncertainty in any extrapolation of the observed conditions beyond the perimetre of the tunnel.

When used in conjunction with nearby subsurface investigation data and geotechnical instrumentation records, the face maps may be used to develop general correlations between ground displacement, geological conditions and other factors (depth of tunnel, groundwater conditions, etc.).

Geotechnical Instrumentation

Geotechnical instrumentation is used during construction to monitor ground and structure displacements, surface settlement above and near the tunnel, deformation of the initial tunnel supports and final lining, groundwater levels, loads in structural elements of the excavation support systems, and ground and structure vibrations, among others. Such instrumentation is a key element of any program for maintenance and protection of existing structures and facilities. In addition, it provides quantitative information for assessing tunnelling procedures during the course of construction, and can be used to trigger modifications to tunnelling procedures in a timely manner to reduce the impacts of construction. Instrumentation is also used to monitor the deformation and stability of the tunnel opening, to assess the adequacy of the initial tunnel support systems and the methods and sequencing of tunnelling, particularly for tunnels constructed by the Sequential Excavation Method (SEM) and tunnels in shear zones or squeezing ground.

Probing

If applicable, such as for SEM and hard rock tunnelling projects, probing ahead of the tunnel face is used to determine general ground conditions in advance of excavation, and to identify and relieve water pressures in any localised zones of water-bearing soils or rock joints. For tunnels constructed by SEM, probing also provides an early indication of the type of ground supports that may be needed as the excavation progresses. Advantages of probing are a) it reduces the risks and hazards associated with tunnelling, b) it provides continuous site investigation data directly along the path of the tunnel, c) it provides information directly ahead of the tunnel excavation, allowing focus on ground conditions of most immediate concern to tunnelling operations, and d) it can be performed quickly, at relatively low cost. However, disadvantages include a) the risk of missing important features by drilling only a limited number of probe holes from the face, and b) the interruption to tunnelling operations during probing. Probing from within the tunnel must be considered as a supplementary investigation method, to be used in conjunction with subsurface investigation data obtained during other phases of the project.

Probing typically consists of drilling horizontally from the tunnel heading by percussion drilling or rotary drilling methods. Coring can be used for probing in rock, but is uncommon due to the greater time needed for coring. Cuttings from the probe holes are visually examined and classified, and assessed for potential impacts to tunnel excavation and support procedures. In rock, borehole cameras can be used to better assess rock quality, orientation of discontinuities, and the presence of shear zones and other important features..

The length of the probe holes can vary considerably, ranging from just 3 or 4 times the length of each excavation stage (round), to hundreds of feet. Shorter holes can be drilled more quickly, allowing them to be performed as part of the normal excavation cycle. However, longer holes, performed less frequently, may result is fewer interruptions to tunnelling operations.

Pilot Tunnels

Pilot tunnels (and shorter exploratory adits) are small size tunnels (typically at least 6.5 ft by 6.5 ft in size) that are occasionally used for large size rock tunnels in complex geological conditions. Pilot tunnels, when used, are typically performed in a separate contract in advance of the main tunnel contract to provide prospective bidders a clearer understanding of the ground conditions that will be encountered. Although pilot tunnels are a very costly method of exploration, they may result in considerable financial benefits to the client by a) producing bids for the main tunnel work that have much lower contingency fees, and b) reducing the number and magnitude of differing site condition claims during construction.

In addition to providing bidders the opportunity to directly observe and assess existing rock conditions, pilot tunnels also offer other significant advantages, including a) more complete and reliable information for design of both initial tunnel supports and final lining, if any, b) access for performing in situ testing of the rock along the proposed tunnel, c) information for specifying and selecting appropriate methods of construction and tunnelling equipment, d) an effective means of pre-draining groundwater, and more confidently determining short-term and long-term groundwater control measures, e) an effective means for identifying and venting gassy ground conditions, e) a means for testing and evaluating potential tunnelling methods and equipment, and f) access for installation of some of the initial supports (typically in the crown area of the tunnel) in advance of the main tunnel excavation. Consideration can also be given to locating the pilot tunnel adjacent to the proposed tunnel, using the pilot tunnel for emergency egress, tunnel drainage, tunnel ventilation, or other purposes for the completed project.

Geospatial Data Management System

Geographic Information System (GIS) is designed for managing a large quantity of data in a complex environment, and is a capable tool used for decision making, planning, design, construction and program management. It can accept all types of data, such as digital, text, graphic, tabular, imagery, etc., and organise these data in a series of interrelated layers for ready recovery. Information stored in the system can be selectively retrieved, compared, overlain on other data, composite with several other data layers, updated, removed, revised, plotted, transmitted, etc.

GIS can provide a means to enter and quickly retrieve a wide range of utility information, including their location, elevation, type, size, date of construction and repair, ownership, right-of-way, etc. This information is stored in dedicated data layers, and can be readily accessed to display or plot both technical and demographic information.

Typical information that could be input to a GIS database for a tunnel project may include street grids; topographic data; property lines; right-of-way limits; existing building locations, type of construction, heights, basement elevations, building condition, etc.; proposed tunnel alignment and profile information; buried abandoned foundations and other underground obstructions; alignment and elevations for existing tunnels; proposed structures, including portals, shafts, ramps, buildings, etc.; utility line layout and elevations, vault locations and depths; boring logs and other subsurface investigation information; geophysical data; inferred surfaces for various soil and rock layers; estimated groundwater surface; areas of identified soil and groundwater contamination; and any other physical elements of jurisdictional boundaries within the vicinity of the project.

Geotechnical Reports

Conventionally, for typical roadway and bridge projects, the geotechnical engineer prepares "geotechnical reports" that serve to summarise the subsurface investigations performed, interpret the existing geological conditions, establish the geotechnical design parameters for the various soil and rock strata encountered, provide geotechnical recommendations for design of the proposed foundations and/or geotechnical features, and identify existing conditions that may influence construction. The term geotechnical report is often used generically to include all types of geotechnical reports, e.g. geotechnical investigation report, geotechnical design report, landslide study report, soil report, foundation report, etc. (FHWA, 1988). The concept is the geotechnical report is only used to communicate the site conditions and design and construction recommendations to the roadway design, bridge design and construction personnel. It may or may not be made available to prospective contractors; and when provided, they are generally only included as a reference document and may typically include disclaimers stating that the report is not intended to be used for construction, and that there is no warranty regarding the accuracy of the data or the conclusions and recommendations of the report; contractors must make their own interpretation of the data to determine the means and associated costs for construction.

Although this approach is commonly used, and may still be applicable for cut-and-cover and immersed tunnel projects, it is not appropriate for mined and bored tunnel and other underground construction projects. Underground projects entail great uncertainty and risk in defining typically complex geological and groundwater conditions, and in predicting ground behaviour during tunnelling operations. Even with extensive subsurface investigations, considerable judgment is required in the interpretation of the subsurface investigation data to establish geotechnical design parameters and to identify the issues of significance for tunnel construction. This situation is further complicated for tunnelling projects since the behaviour of the ground during construction is typically influenced by the contractor's selected means and methods for tunnel excavation and type and installation of tunnel supports.

Using conventional geotechnical reports for tunnel projects would essentially assign the full risk of construction to the contractor since the contractor is responsible for interpreting the available subsurface information. Although this approach appears to protect the owner from the uncertainties and risks of construction, experience on underground projects has demonstrated that it results in high contingency costs being included in the contractors' bids, and does not avoid costly contractor claims for additional compensation when subsurface conditions vary from those that could reasonably be anticipated.

Current practice for tunnel and underground projects in the U.S. seeks to obtain a more equitable sharing of risks between the contractor and the owner. This approach recognises that owners largely define the location, components and requirements of a project and the extent of the site investigations performed, and therefore should accept some of the financial risk should ground conditions encountered during construction differ significantly from those anticipated during design and preparation of the contract documents, and should they negatively impact the contractor. The overall objectives of this risk sharing approach are to:

- Reduce the contractors' uncertainty regarding the financial risks of tunnelling projects to obtain lower bid prices
- Foster greater cooperation between the contractor and the owner
- Quickly and equitably resolve disputes between the contractor and the owner that may arise when ground conditions encountered during construction differ substantially from those reflected in the contract documents at the time of bidding
- Obtain the lowest final cost for the project.

Contracting practices for underground projects in the U.S. have evolved and currently include a number of measures to help achieve the above objectives. These measures vary somewhat between projects, depending on specific project conditions and owner preferences, but typically consist of the following fundamental elements:

- Thorough geotechnical site investigations
- Full disclosure of available geotechnical information to bidding contractors
- Preparation of a Geotechnical Data Report (GDR) to present all the factual data for a project

- Preparation of Geotechnical Design Memorandum (GDM) to present an interpretation of the available geotechnical information, document the assumptions and procedures used to develop the design, and facilitate communication within the design team during development of the design. GDMs are not intended to be incorporated into the Contract Documents and are subsequently superseded by the Geotechnical Baseline Report (GBR).
- Preparation of a Geotechnical Baseline Report (GBR) to define the baseline conditions on which contractors will base their bids and select their means, methods and equipment, and that will be used as a basis for determining the merits of contractor claims of differing site conditions during construction
- Making the GDR and GBR contractually binding documents by incorporating them within the contract documents for the project; the GBR takes precedence;
- Carefully coordinating the provisions of the contract specifications and drawings with the information presented in the GBR
- Including a Differing Site Condition clause in the specification that allows the contractor to seek compensation when ground conditions vary from those defined in the GBR, and that result in a corresponding increase in construction cost and/or delay in the construction schedule; Establishing a dispute resolution process to quickly and equitably resolve disagreements that may arise during construction without reverting to costly litigation procedures
- Providing escrow of bid documents

Geotechnical Data Report

The Geotechnical Data Report (GDR) is a document that presents the factual subsurface data for the project without including an interpretation of these data. The purpose of the GDR is to compile all factual geological, geotechnical, groundwater, and other data obtained from the geotechnical investigations for use by the various participants in the project, including the owner, designers, contractors and third parties that may be impacted by the project. It serves as a single and comprehensive source of geotechnical information obtained for the project.

The GDR should avoid making any interpretation of the data since these interpretations may conflict with the data assessment subsequently presented in the Geotechnical Design Memorandum or other geotechnical interpretive or design reports, and the baseline conditions defined in the Geotechnical Baseline Report. Any such discrepancies could be a source of confusion to the contractors and open opportunities for claims of differing site conditions.

In practice, it may not be possible to eliminate all data interpretation from the GDR. In such case, the data reduction should be limited to a determination of the properties obtained from that individual test sample, while avoiding any recommendations for the geotechnical properties for the stratum from which the sample was obtained.

The GDR should contain the following information (ASCE, 2007):

- Descriptions of the geologic setting
- Descriptions of the site exploration program(s)
- Logs of all borings, trenches, and other site investigations
- Descriptions/discussions of all field and laboratory test programs
- Results of all field and laboratory testing

The GDR would include the logs of all borings performed for the project, but should not present a subsurface profile constructed from the borings since such a profile requires considerable judgment and interpolation of the borehole records to show inferred strata boundaries. As illustrated in the outline, the text of the GDR provides background information and a discussion of the subsurface investigations performed, while the specific data are presented in appendices to the report.

The introduction provides a general project description and notes the purpose and scope of the report. The section on background information should identify other sources of geotechnical information that may have been obtained by others at or near the project site, and may include subsurface investigation data and records from previous construction activities. If such additional information is limited in volume, consideration should be given to including these data in an appendix to the report.

Background information should also include a discussion of the regional and local geologic setting, since such information will be invaluable in the assessment of the limited amount of factual data obtained from site investigations. It is recognised that a description of geological conditions requires interpretation of information in the literature and an understanding of the geological processes controlling the formation and properties of soil and rock deposits; however, since an understanding of the geological setting is fundamental to a successful tunnelling project, such information is considered an essential component of the GDR.

Geotechnical Design Memorandum

For tunnel projects, one or more interpretive reports may be prepared to evaluate the available data as presented in the GDR, address a broad range of design issues, and communicate design recommendations for the design team's internal consideration. These interpretive reports are also used to evaluate design alternatives, assess the impact of construction on adjacent structures and facilities, focus on individual elements of the project, and discuss construction issues. The current guidelines recommend referring to such design reports as Geotechnical Design Memoranda (GDM), instead of Geotechnical Interpretive Report (GIR) (ASCE, 2007).

GDM, or GIR may be prepared at different stages of a project, and therefore may not accurately reflect the final design or final contract documents. Hence, preparation of such interpretive reports in the course of the final design is superfluous, and is strongly discouraged to avoid a potential source of confusion and conflict.

Since GDMs are used internally within the design team and with the owner as part of the project development effort, it is not appropriate to include GDMs as part of the contract documents. Thus, GDMs should be clearly differentiated from the Geotechnical Baseline Report (GBR). The GBR should be the only interpretive report prepared for use in bidding and constructing the project. The GBR must supersede – take precedence over any other Geotechnical Report(s). However, in the interest of "full disclosure" to prospective bidders, GDMs are often made available "for information only." In such instances, the GDM must include a disclaimer clearly noting the specific purposes of the report and stating that the information provided in the report is not intended for construction. The GDM must also clearly state that the contract documents, including the GDR and GBR, are the only documents to be considered by contractors when assessing project requirements and determining their bid price for the work.

The GDM should include other disclaimers to highlight the interpretive nature of the report. Following are several issues that are commonly addressed by disclaimers:

- The boring logs only represent the conditions at the specific borehole location at the time it was drilled; ground conditions may be different beyond the borehole location, and may change with time as a result of nearby activities as well as natural processes
- Water levels in the boreholes and observation wells are seasonal and may also change as a result of other factors
- The findings and recommendations presented in the report are applicable only to the proposed facilities and should not be used for other purposes

In evaluating the engineering properties of the soil and rock materials, it is appropriate for the GDM to note the likely ranges for these properties and to recommend a value, or range of values, for use in design. The report should document the basis for selecting these parameters and discuss their significance to the design and construction of the proposed facilities.

Presenting a range of parameters, along with a discussion of their consequences on the design, helps the owner and the design team understand and quantify the inherent uncertainty and risk associated with the proposed underground project. Such information allows the owner to determine the level of risk to be accepted, and the share of the risk to be borne by the contractor. An example of this decision process would be a case where a tunnel must be constructed through relatively low strength rock that contains intrusive dikes of very hard igneous rock of unknown frequency and thickness.

Based on limited geotechnical investigations, the geotechnical engineer determines that the amount of hard rock may range from 10 to 30 percent of the total length of the tunnel. This range, and possibly a best estimate percentage, would be reported in the GDM. During subsequent preparation of the GBR and other contract documents, a specific baseline value would be determined and referenced for contractual purposes and reflected in the design. If the owner, in an effort to get lower bid prices, is willing to accept the greater risk of cost increases during construction, a value closer to the lower end of the range would be selected as the baseline. However, if the owner wishes to reduce the risk of cost extras during construction, a value closer to the conservative end of the range would be selected. However, in choosing this second option, the owner needs to recognise that it will result in higher bid prices.

In preparing a GDM it is acceptable to use ambiguous terms, such as "may," "should," "likely," etc. in discussing the various technical issues. Such terms reflect the reality of our uncertainty in defining subsurface stratigraphy and the engineering properties of natural materials, and in predicting the behaviour of these materials during construction.

The GDM should reference the Geotechnical Data Report (GDR) as the source of information used to develop the conclusions and recommendations of the GDM. The GDM should also identify any other sources of information that may have influenced the findings of the GDM, including technical references, reports, and site reconnaissance observations, among others.

The GDM should include generalised subsurface profiles developed from an assessment of the available geotechnical and geological information. These subsurface profiles greatly facilitate a visualization and understanding of the existing subsurface conditions for design purposes. However, it must be recognised that such definition of subsurface conditions is highly dependent on the quantity and quality of the available geotechnical investigation data, and the judgment of the geotechnical engineer in interpreting these data and the relevant geological information. Accordingly, the report must emphasize that the profiles are based on an interpolation between widely spaced borings, and that actual subsurface conditions between the borings may vary considerably from those indicated on the profiles.

In addition to providing recommendations for design, the GDM should also address construction issues, including the general methods of construction considered appropriate for the existing site conditions and proposed facilities. However, the engineers preparing the report must recognise that the contractor is responsible for selecting the specific equipment, means and methods for performing the work, thus must avoid any detailed recommendations on these issues accordingly. For example, for a proposed subaqueous tunnel through highly permeable sand deposits, it is appropriate to state that a closed face Tunnel Boring Machine (TBM) consisting of either an Earth Pressure Balance (EPB) Shield or Slurry Shield TBM should be used, but it is inappropriate to recommend a specific TBM model, horsepower, etc.

It is also particularly important for the GDM to identify and discuss all potential hazards that may be encountered during construction, and discuss possible measures to mitigate these hazards. A thorough discussion of such issues should help both the design team and the contractor to anticipate and avoid problems that could cause major cost and schedule impacts. For example, for a tunnel to be excavated in mixed face conditions, the GDM should note that typical problems may include: a), large water inflow at the contact between the soil and rock that will be difficult to fully dewater, b), steering problems for the TBM, and c), ground loss, and corresponding surface settlement due to excavation of the soil in the upper part of the tunnel heading at a faster rate than the rock in the lower part of the heading. For this example, the report should also note mitigating measures, such as grouting the soil to reduce seepage and ground loss, and facilitate steering of the TBM; drilling drainage holes horizontally from the tunnel heading; providing an articulated TBM to facilitate steerage corrections; etc.

In summary, the GDM is written by the engineers solely for use by the design team in developing the design for the proposed facilities. It provides an interpretation of the available subsurface information to determine likely subsurface conditions for design purposes. Depending on its specific purpose and the time of its preparation, the GDM may not reflect the final design shown on the contract drawings. An important element of the GDM is a general discussion of the appropriate methods of construction and the potential hazards that may be encountered during construction, as well as the possible measures that can be considered to mitigate these hazards. The GDM is not intended to be a definitive representation of the actual ground conditions, and is not to be used as a baseline for contractual purposes.

Geotechnical Baseline Report

Purpose and Objective: A fundamental principal in current U.S. contracting practices for tunnel projects is the equitable sharing of risk between the owner and contractor, with the objectives of reducing contingency fees in contractor bids, achieving lower total cost for the project, and streamlining resolution of contractor claims for changed conditions during construction. Over the years, various forms and names have been given to the interpretive geotechnical report to be incorporated into the Contract Documents for underground projects in order to achieve the aforementioned objectives. Originally, this was called the Geotechnical Design Summary Report (GDSR). However, since 1997 and continuing with the current (2007) "Geotechnical Baseline Reports-Suggested Guidelines" the industry has determined that the incorporated report be called GBR (Geotechnical Baseline Report).

The primary purposes of the GBR are:

- Establish a contractual document that defines the specific subsurface conditions to be considered by contractors as baseline conditions in preparing their bids,
- Establish a contractual procedure for cost adjustments when ground conditions exposed during construction are poorer than the baseline conditions defined in the contract documents.

Although it reflects the findings of the geotechnical investigations and design studies, a GBR is not intended to predict the actual geotechnical and geological conditions at a project site, or to accurately predict the ground behaviour during construction. Rather, it establishes the bases for delineating the financial risks between the owner and the contractor.

ASCE (1997) also noted the secondary purposes of the GBR as listed below:

- It presents the geotechnical and construction considerations that formed the basis of design
- It enhances contractor understanding of the key project issues and constraints, and the requirements of the contract plans and specifications
- It identifies important considerations that need to be addressed during bid preparation and construction
- It assists the contractor in evaluating the requirements for tunnel excavation and support; and
- It guides the construction manager in administering the contract and monitoring contractor performance

A common misconception of the GBR is that it represents a warranty of the existing site conditions by the geotechnical engineer and designer. Based on this understanding, the owner of the project may believe they are entitled to compensation by the designer should actual conditions be found less favourable than the conditions defined in the GBR. However, since it principally serves as a contractual instrument for allocating risks, the GBR is not intended to predict or warranty actual site conditions. If the GBR were to become a warranty, it is reasonable to expect that the geotechnical engineer and designer would more conservatively define subsurface conditions and ground behaviour, resulting in a higher cost for the project, a consequence clearly contrary to the primary motivation for adopting a risk-sharing approach to tunnel construction contracts.

It is also important to clearly differentiate the GBR from other interpretive reports may be prepared by the design team to addressing a broad range of design issues for the team's internal consideration. Such reports should be referred to as Geotechnical Design Memoranda (GDM). The GBR should be the only final report prepared for use in bidding and constructing the project. The GBR should be limited to interpretive discussion and baseline statements, and should make reference to, rather than repeat or paraphrase, information contained in the GDR, drawings, or specifications (ASCE, 2007).

General Considerations

The various elements of the construction contract documents each serve a different purpose. The GDR provides the factual information used by the designer for designing the various components of the project, and by the contractor for developing appropriate means and methods of construction. The contract plans and specifications detail the specific requirements for the work to be performed, without providing an explanation or background information. The GBR is based on the factual information presented in the GDR as well as input from the owner regarding risk allocation, and provides an explanation for the project requirements as presented in the contract plans and specifications. The baseline information presented in the GBR must be coordinated with the GDR, contract plans and specifications, and contract payment provisions to assure consistency throughout the contract. However, the GBR should not repeat or paraphrase statements made in these other contract documents since even minor rewording of a statement may cause confusion or an unintended interpretation of the statement. Any inconsistency or confusion in the contract documents could lead to a successful contractor claim for additional compensation during construction since these are usually judged against the owner as the originator of the contract.

The contract General Provisions or Special Provisions should clearly define the hierarchy between the various parts of the contract documents to help resolve any conflicts that may inadvertently remain after issuing the documents. The GBR takes precedence over the GDR and any and all other geotechnical report prepared for any reason.

Most often, there is a possible baseline range that can be established for a given set of geologic and construction conditions. As a consequence, where the baseline is set determines the risk allocation for the project. When an adverse baseline is adopted; 1), more risk is assigned to the Contractor who will bid higher, 2), less risk and reduced potential for change orders accrue to the owner, and, 3), higher costs accrue to the owner due to paying for the contingency of encountering the adverse condition(s). Conversely, when a less adverse baseline is adopted: 1) Contractor bids less due to less risk and contingency, 2) higher risk and potential for change orders accrue to the owner, 3) owner pays more if adverse conditions are encountered but less if they are not encountered. In either case, the cost of site conditions remains with the Owner.

Guidelines for Preparing a GBR

The GBR translates facts, interpretations and opinions regarding subsurface conditions into clear, unambiguous statements for contractual purposes. Items typically addressed in a GBR include:

- The amounts and distribution of different materials along the selected alignment;
- Description, strength, compressibility, grain size, and permeability of the existing materials;
- Description, strength and permeability of the ground mass as a whole;
- Groundwater levels and expected groundwater conditions, including baseline estimates of inflows and pumping rates;
- Anticipated ground behaviour, and the influence of groundwater, with regard to methods of excavation and installation of ground support;
- Construction impacts on adjacent facilities; and
- Potential geotechnical and man-made sources of potential difficulty or hazard that could impact construction, including the presence of faults, gas, boulders, solution cavities, existing foundation piles, and the like.

A general checklist for a GBR is presented in Table 4-3. This checklist assumes that the Geotechnical Data Report contains the information noted.

Following are general guidelines that should be followed for preparation of a Geotechnical Baseline Report:

- The GBR should be brief. The length of a GBR should be limited to not more than 30 pages of text for typical projects, and not more than 50 pages for more complex projects. The length should allow reading the GBR in a single sitting.

- Select baseline parameters following discussions with the owner regarding the levels of risk to be allotted to the owner and contractor
- Use and reference the information presented in the GDR as the basis for selecting baseline parameters
- Avoid using ambiguous terminology, such as "may," "should," "can," etc; rather, use definitive terms, such as "is," "are," "will," etc.
- Whenever possible, refer baselines to properties and parameters that can be objectively observed and measured in the field
- Avoid the use of general adjectives, such as "large," "significant," "minor," etc. unless these terms are defined and quantified
- Carefully select the specific wording used in the GBR to avoid unintended interpretation of the report
- For parameters that are anticipated to vary considerably, the GBR should note the potential range of values, but clearly state a specific baseline value for contractual purposes
- Since ground behaviour is largely influenced by construction means and methods, statements of ground behaviour in the GBR should also note the corresponding construction equipment, procedures and sequencing on which these statements were based
- Include an independent review of the GBR at different stages of completion to identify possible ambiguity and inconsistencies, and to verify that all relevant issues are appropriately addressed.

Individuals who prepare the GBR must be highly knowledgeable of both the design and construction of underground facilities, with construction experience particularly important for the necessary understanding of construction methods, equipment capabilities, ground behaviour during tunnel excavation, and the potential hazards associated with the different ground conditions and methods of construction. In addition, these individuals must be experienced in the preparation of a GBR and clearly understand its role as a contract document establishing reference baseline conditions. In general, to achieve greater consistency in the contract documents, the individuals preparing the GBR should belong to the same organisation that prepares the contract plans and specifications.

Previous Construction Experience (Key Points only in GBR if Detailed in GDR):

- Nearby relevant projects
- Relevant features of past projects, with focus on excavation methods, ground behaviour, groundwater conditions, and ground support methods
- Summary of problems during construction and how they were overcome (with qualifiers as appropriate)

Ground Characterisation:

- Physical characteristic and occurrences of each distinguishable rock or soil unit, including fill, natural soils, and bedrock; describe degree of weathering / alteration; including near-surface units for foundations/pipelines
- Groundwater conditions; depth to water table; perched water; confined aquifers and hydrostatic pressures; pH; and other key groundwater chemistry details
- Soil/rock and groundwater contamination and disposal requirements
- Laboratory and field test results presented in histogram (or some other suitable) format, grouped according to each pertinent distinguishable rock or soil unit; reference to tabular summaries contained in the GDR
- Ranges and values for baseline purposes; explanations for why the histogram distributions (or other presentations) should be considered representative of the range of properties to be encountered, and if not, why not; rationale for selecting the baseline values and ranges
- Blow count data, including correlation factors used to adjust blow counts to Standard Penetration Test (SPT) values, if applicable
- Presence of boulders and other obstructions; baselines for number, frequency (i.e., random or concentrated along geologic contacts), size and strength
- Bulking/swell factors and soil compaction factors
- Baseline descriptions of the depths/thicknesses or various lengths or percentages of each pertinent distinguishable ground type or stratum to be encountered during excavation; properties

of each ground type; cross-references to information contained in the drawings or specifications

- Values of ground mass permeability, including direct and indirect measurements of permeability values, with reference to tabular summaries contained in the GDR; basis for any potential occurrence of large localised inflows not indicated by ground mass permeability values
- For TBM projects, interpretations of rock mass properties that will be relevant to boreability and cutter wear estimates for each of the distinguishable rock types, including test results that might affect their performance (avoid explicit penetration rate estimated or advance rate estimates)

Design Considerations – Tunnels and Shafts:

- Description of ground classification system(s) utilised for design purposes, including ground behaviour nomenclature
- Criteria and methodologies used for the design of ground support and ground stabilisation systems, including ground loadings (or reference the drawings/specifications)
- Criteria and bases for design of final linings (or reference to drawings/specifications)
- Environmental performance considerations such as limitations on settlement and lowering of groundwater levels (or in specifications)
- The manner in which different support requirements have been developed for different ground types, and, if required, the protocol to be followed in the field for determination of ground support types for payment; reference to specifications for detailed descriptions of ground support methods/sequences
- The rationale for ground performance instrumentation included in the drawings and specifications

Design Considerations – Other Excavations and Foundations:

- Criteria and methodologies used for the design of excavation support systems, including lateral earth pressure diagrams (or in drawings/specifications) and need to control deflections/deformations
- Feasible excavation support systems
- Minimum pile tip elevations for deep foundations
- Refusal criteria for driven piles

- Allowable skin friction for tiebacks
- Environmental considerations such as limitations on settlement and lowering of groundwater levels (or in specifications)
- Rationale for instrumentation/monitoring shown in the drawings and specifications

Construction Considerations – Tunnels and Shafts:

- Anticipated ground behaviour in response to construction operations within each soil and rock unit
- Required sequences of constructions (or in drawings/ specifications)
- Specific anticipated construction difficulties
- Rationale for requirements contained in the specifications that either constrain means and methods considered by the contractor or prescribe specific means and methods, e.g., the required use of an EPB or slurry shield.
- The rationale for baseline estimates of groundwater inflows to be encountered during construction, with baselines for sustained inflows at the heading, flush inflows at the heading, and cumulative sustained groundwater inflows to be pumped at the portal or shaft
- The rationale behind ground improvement techniques and groundwater control methods included in the contract
- Potential sources of delay, such as groundwater inflows, shears and faults, boulders, logs, tiebacks, buried utilities, other manmade obstruction, gases, contaminated soils and groundwater, hot water, and hot rock, etc.

Construction Considerations – Other Excavations and Foundations:

- Anticipated ground behaviour in response to required construction operations within each soil and rock unit
- Rippability of rock, till, caliche, or other hard materials, and other excavation considerations including blasting requirements/limitations
- Need for groundwater control and feasible groundwater control methods
- Casing requirements for drilled shafts
- Specific anticipated construction difficulties

- Rationale for requirements contained in the specifications that either constrain means and methods considered by the Contractor or prescribe specific means and methods
- The rationale for baseline estimates of groundwater inflows to be encountered during construction, with baselines for sustained inflows to be pumped from the excavation
- The rationale behind ground improvement techniques and groundwater control methods included in the Contract
- Potential sources of delay, such as groundwater inflows, shears and faults, boulders, buried utilities, manmade obstruction, gases, or contaminated soils or groundwater

Cut and Cover Tunnels

This chapter presents the construction methodology and excavation support systems for cut-and-cover road tunnels and describes the structural design in accordance with the AASHTO LRFD Bridge Design Specifications (AASHTO, 2008). The intent of this chapter is to provide guidance in the interpretation of the AASHTO LRFD Specifications in order to have a more uniform application of the code and to provide guidance in the design of items not specifically addressed in AASHTO (2008). The designers must follow the latest LRFD Specifications. A design example illustrating the concepts presented in this chapter can be found in Appendix C. Other considerations dealing with support of excavation, maintenance of traffic and utilities, and control of groundwater and how they affect the structural design are discussed.

Construction Methodology

In a cut and cover tunnel, the structure is built inside an excavation and covered over with backfill material when construction of the structure is complete. Cut and cover construction is used when the tunnel profile is shallow and the excavation from the surface is possible, economical, and acceptable. Cut and cover construction is used for underpasses, the approach sections to mined tunnels and for tunnels in flat terrain or where it is advantageous to construct the tunnel at a shallow depth. Two types of construction are employed to build cut and cover tunnels; bottom-up and top-down. These construction types are described in more detail below. The planning process used to determine the appropriate profile and alignment for tunnels.

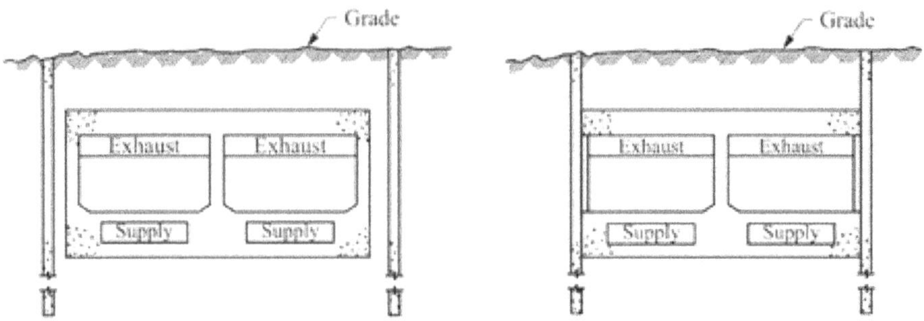

Figure: Cut and Cover Tunnel Bottom-Up Construction (a); Top-Down Construction (b)

For depths of 30 to 40 feet (about 10 m to 12 m), cut-and-cover is usually more economical and more practical than mined or bored tunnelling. The cut-and-cover tunnel is usually designed as a rigid frame box structure. In urban areas, due to the limited available space, the tunnel is usually constructed within a neat excavation line using braced or tied back excavation supporting walls. Wherever construction space permits, in open areas beyond urban development, it may be more economical to employ open cut construction.

Where the tunnel alignment is beneath a city street, the cut-and-cover construction will cause interference with traffic and other urban activities. This disruption can be lessened through the use of decking over the excavation to restore traffic. While most cut-and-cover tunnels have a relatively shallow depth to the invert, depths to 60 feet (18 m) are not uncommon; depths rarely exceed 100 feet (30 m).

Conventional Bottom-Up Construction

As shown, in the conventional "bottom-up" construction, a trench is excavated from the surface within which the tunnel is constructed and then the trench is backfilled and the surface restored afterward. The trench can be formed using open cut (sides sloped back and unsupported), or with vertical faces using an excavation support system. In bottom-up construction, the tunnel is completed before it is covered up and the surface reinstated.

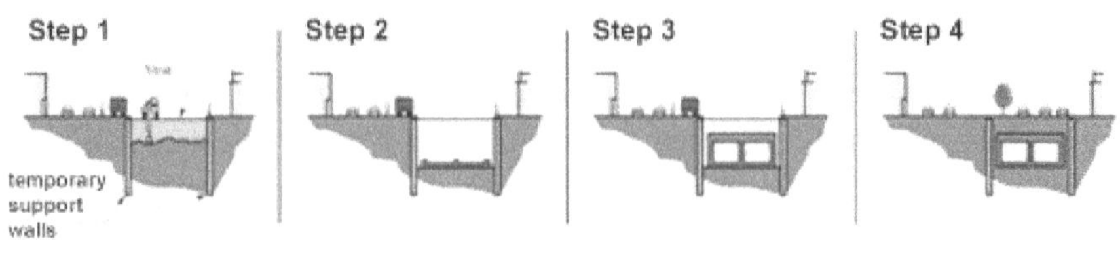

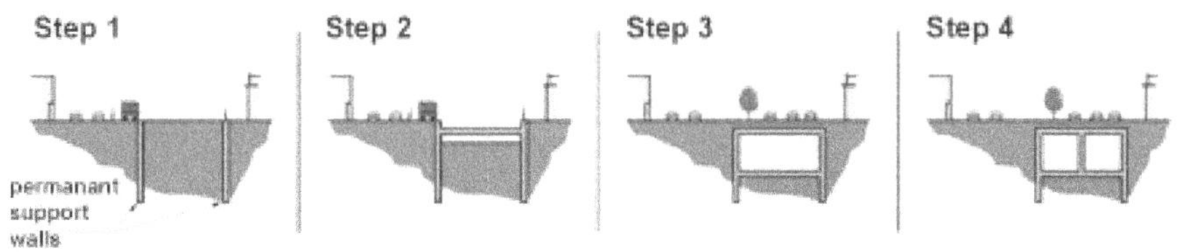

Figure: Cut-and-Cover Tunnel Bottom-Up (a) and Top-Down (b) Construction Sequence

Conventional bottom-up sequence of construction in Figure mentioned generally consists of the following steps:

- Step 1a: Installation of temporary excavation support walls, such as soldier pile and lagging, sheet piling, slurry walls, tangent or secant pile walls
- Step 1b: Dewatering within the trench if required
- Step 1c: Excavation and installation of temporary wall support elements such as struts or tie backs
- Step 2: Construction of the tunnel structure by constructing the floor;
- Step 3: Compete construction of the walls and then the roof, apply waterproofing as required;
- Step 4: Backfilling to final grade and restoring the ground surface.

Bottom-up construction offers several advantages:

- It is a conventional construction method well understood by contractors.
- Waterproofing can be applied to the outside surface of the structure.
- The inside of the excavation is easily accessible for the construction equipment and the delivery, storage and placement of materials.
- Drainage systems can be installed outside the structure to channel water or divert it away from the structure.

Disadvantages of bottom-up construction include:

- Somewhat larger footprint required for construction than for top-down construction.
- The ground surface can not be restored to its final condition until construction is complete.

- Requires temporary support or relocation of utilities.
- May require dewatering that could have adverse affects on surrounding infrastructure.

Top-Down Construction

With top-down construction, the tunnel walls are constructed first, usually using slurry walls, although secant pile walls are also used. In this method the support of excavation is often the final structural tunnel walls. Secondary finishing walls are provided upon completion of the construction. Next the roof is constructed and tied into the support of excavation walls. The surface is then reinstated before the completion of the construction. The remainder of the excavation is completed under the protection of the top slab. Upon the completion of the excavation, the floor is completed and tied into the walls. The tunnel finishes are installed within the completed structure. For wider tunnels, temporary or permanent piles or wall elements are sometimes installed along the centre of the proposed tunnel to reduce the span of the roof and floors of the tunnel.

Top-down sequence of construction generally consists of the following steps:

- Step 1a : Installation of excavation support/tunnel structural walls, such as slurry walls or secant pile walls
- Step 1b: Dewatering within the excavation limits if required
- Step 2a: Excavation to the level of the bottom of the tunnel top slab
- Step 2b: Construction and waterproofing of the tunnel top slab tying it to the support of excavation walls
- Step 3a: Backfilling the roof and restoring the ground surface
- Step 3b: Excavation of tunnel interior, bracing of the support of excavation walls is installed as required during excavation
- Step 3c: Construction of the tunnel floor slab and tying it to the support of excavation walls; and
- Step 4 Completing the interior finishes including the secondary walls.

Top-down construction offers several advantages in comparison to bottom-up construction:

- It allows early restoration of the ground surface above the tunnel

- The temporary support of excavation walls are used as the permanent structural walls
- The structural slabs will act as internal bracing for the support of excavation thus reducing the amount of tie backs required
- It requires somewhat less width for the construction area
- Easier construction of roof since it can be cast on prepared grade rather than using bottom forms
- It may result in lower cost for the tunnel by the elimination of the separate, cast-in-place concrete walls within the excavation and reducing the need for tie backs and internal bracing
- It may result in shorter construction duration by overlapping construction activities

Disadvantages of top-down construction include:

- Inability to install external waterproofing outside the tunnel walls.
- More complicated connections for the roof, floor and base slabs.
- Potential water leakage at the joints between the slabs and the walls
- Risks that the exterior walls (or centre columns) will exceed specified installation tolerances and extend within the neat line of the interior space.
- Access to the excavation is limited to the portals or through shafts through the roof.
- Limited spaces for excavation and construction of the bottom slab

Selection

It is difficult to generalise the use of a particular construction method since each project is unique and has any number of constraints and variables that should be evaluated when selecting a construction method. The following summary presents conditions that may make a one construction method more attractive than the other. This summary should be used in conjunction with a careful evaluation of all factors associated with a project to make a final determination of the construction method to be used.

Conditions Favourable to Bottom-Up Construction:

- No right-of way restrictions
- No requirement to limit sidewall deflections

- No requirement for permanent restoration of surface

Conditions Favourable to Top-Down Construction

- Limited width of right-of-way
- Sidewall deflections must be limited to protect adjacent features
- Surface must be restored to permanent usable condition as soon as possible

Support of Excavation

The practical range of depth for cut and cover construction is between 30 and 40 feet (about 10 m to 12 m). Sometimes, it can approach 100 feet.

Excavations for building cut and cover tunnels must be designed and constructed to provide a safe working space, provide access for construction activities and protect structures, utilities and other infrastructure adjacent to the excavation. The design of excavation support systems requires consideration of a variety of factors that affect the performance of the support system and that have impacts on the tunnel structure itself. These factors are discussed hereafter.

Excavation support systems fall into three general categories:

- *Open cut slope:* This is used in areas where sufficient room is available to open cut the area of the tunnel and slope the sides back to meet the adjacent existing ground line. The slopes are designed similar to any other cut slope taking into account the natural repose angle of the in-situ material and the global stability.
- *Temporary:* This is a structure designed to support vertical or near vertical faces of the excavation in areas where room to open cut does not exist. This structure does not contribute to the final load carrying capacity of the tunnel structure and is either abandoned in placed or dismantled as the excavation is being backfilled. Generally it consists of soldier piles and lagging, sheet pile walls, slurry walls, secant piles or tangent piles.
- *Permanent:* This is a structure designed to support vertical or near vertical faces of the excavation in areas where room to open cut does not exist. This structure forms part of the permanent final tunnel structure. Generally it consists of slurry walls, secant pile walls, or tangent pile walls.

This section discusses temporary and permanent support of excavation systems and provides issues and concerns that must be considered during the development of a support of excavation scheme. The design of open-cut slopes and support of excavation are not in the scope of this manual. Information on the design of soil and rock slopes can be found in FHWA-NHI-05-123 "Soil Slope and Embankment Design" (FHWA, 2005d), and NHI-99-007 "Rock Slopes" (FHWA, 1999), respectively. Supports of Excavation are referred to FHWA-NHI-05-046 "Earth Retaining Structure" (FHWA, 2005e). Many of the issues described below associated with ground and groundwater behaviour are applicable to side slopes also.

Figure: Cut and Cover Construction using Side Slopes Excavation- Ft McHenry Tunnel, Baltimore, MD

Temporary Support of Excavation

Support of excavation structures can be classified as flexible or rigid. Flexible supports of excavation include sheet piling and soldier pile and lagging walls. A careful site investigation that provides a clear understanding of the subsurface conditions is essential to determining the correct support system. Rigid support of excavation such as slurry walls, secant piles or tangent piles are also used as temporary support of excavation.

A sheet piling wall consists of a series of interlocking sheets that form a corrugated pattern in the plan view of the wall. The sheets are either driven or vibrated into the ground. The sheets extend well below the bottom of the excavation for stability. These sheets are fairly flexible and can support only small heights of earth without bracing. As the excavation progresses, bracings or tie backs are installed at specified intervals. Sheet pile walls can be installed quickly and easily in ideal soil conditions. The presence of rock, boulders, debris, utilities, or obstructions will make the use of sheet piling difficult since these features will either damage the sheet pile or in the case of a utility, be damaged by the sheet pile.

Figure: Sheet Pile Walls with Multi Level-Bracing

A soldier pile wall consists of structural steel shape columns spaced 4 to 8 feet apart and driven into the ground or placed in predrilled holes. The soldier piles extend well below the level of the bottom of excavation for stability. As the excavation progresses, lagging is placed between the soldier piles to retain the earth behind the wall. The lagging could be timber or concrete planks. The soldier piles are relatively flexible and are capable of supporting only modest heights of earth without bracing. As the excavation progresses, bracing or tie backs are installed at specified intervals. Soldier piles can also be installed in more different ground conditions than can a sheet pile wall. The spacing allows the installation of piles around utilities. The finite dimension of the pile allows drilling of holes through obstructions and into rock, making the soldier pile and lagging wall more versatile than the sheet pile wall.

Figure: Braced Soldier Pile and Lagging Wall

Support of excavation bracing can consist of struts across the excavation to the opposite wall, knee braces that brace the wall against the ground, and tie backs consisting of rock anchors or soil anchors that tie the wall back into the earth behind the wall. Struts and braces extend into the working area and create obstacles to the construction of the tunnel. Tie backs do not obstruct the excavation space but sometimes they extend outside of the available right-of-way requiring temporary underground easements. They may also encounter obstacles such as boulders, utilities or building foundations. The suitability of tie backs depends on the soil conditions behind the wall. The site conditions must be studied and understood and taken into account when deciding on the appropriate bracing method.

The design and detailing of the support of excavation must consider the sequence of installation and account for the changing loading conditions that will occur as the system is installed. The design of temporary support of excavation is not in the scope of this manual. The information presented herein is intended to make tunnel designers aware of the impact that the selected support of excavation can have on the design, constructability and serviceability of the tunnel structure. Guidance on the design of support of excavation can be found in FHWA-NHI-05-046 "Earth Retaining Structure" (FHWA, 2005e).

Figure: Tie-back Excavation Support leaves Clear Access

Use of temporary support of excavation does have the advantage of allowing waterproofing to be applied to the outside face of the tunnel structure. This can be accomplished by setting the face of the support of excavation away from the outside face of the tunnel structure. This space provides room for forming and allows the placement of waterproofing directly onto the finished outside face of the structure. As an alternate, the face of the support of excavation can be placed directly adjacent to the outside face of the structure. Under this scenario, the face of the support of excavation is used as the form for the tunnel structure. Waterproofing is installed against the support of excavation and concrete is poured against the waterproofing. In this case, the temporary support of excavation wall is abandoned in place.

Permanent Support of Excavation

Permanent support of excavation typically employs rigid systems. Rigid systems consist of slurry walls, soldier pile tremie concrete (SPTC) walls, tangent pile walls, or secant pile walls. As with temporary support of excavation systems, a careful site investigation that provides a clear understanding of the subsurface conditions is essential to determining the appropriate system.

A slurry wall is constructed by excavating a trench to the thickness required for the external structural wall of the tunnel. Slurry walls are usually 30 to 48 inches thick. The trench is kept open by the placement of bentonite slurry in the trench as it is excavated. The trench will typically extend for some distance below the bottom of the tunnel structure for stability. Reinforcing steel is lowered into the slurry filled trench and concrete is then placed using the tremie method into the trench displacing the slurry. The resulting wall will eventually be incorporated into the final tunnel structure. Excavation proceeds from the original ground surface down to the bottom of the roof of the tunnel structure. The tunnel roof is constructed and tied into the slurry wall. The tunnel roof provides bracing for the slurry wall. Depending on the depth of the tunnel, the roof could be the first level of bracing or an intermediate level. The excavation would then proceed and additional bracing would be provided as needed. At the base of the excavation, the tunnel bottom slab is then constructed and tied into the walls.

SPTC walls are constructed in the same sequence as a slurry wall. However, once the trench is excavated, steel beams or girders are lowered into the slurry in addition to reinforcing steel to provide added capacity. The construction of the wall then follows the same sequence as that described above for a slurry wall.

Tangent pile (drilled shaft) walls consist of a series of drilled shafts located such that the adjacent shafts touch each other, hence the name tangent wall. The shafts are usually 24 to 48 inches in diametre and extend below the bottom of the tunnel structure for stability. The typical sequence of construction of tangent piles begins with the excavation of every third drilled shaft. The shafts are held open if required by temporary casing. A steel beam or reinforcing bar cage is placed inside the shaft and the shaft is then filled with concrete. If a casing is used, it is pulled as the tremie concrete placement progresses. Once the concrete backfill cures sufficiently, the next set of every third shaft is constructed in the same sequence as the first set. Finally, after curing of the concrete in the second set, the third and final set of shafts is constructed, completing the walls. Excavation within the walls then proceeds with bracing installed as required to the bottom of the excavation. Roof and floor slabs are constructed and tied into the tangent pile. The roof and floor slabs act as bracing levels.

Figure: Tangent Pile Wall Support

Secant pile walls are similar to tangent pile walls except that the drilled shafts overlap each other rather than touch each other. This occurs because the centre to centre spacing of secant piles is less than the diametre of the piles. Secant pile walls are stiffer than tangent piles walls and are more effective in keeping ground water out of the excavation. They are constructed in the same sequence as tangent pile walls. However, the installation of adjacent secant piles requires the removal of a portion of the previously constructed pile, specifically a portion of the concrete backfill.

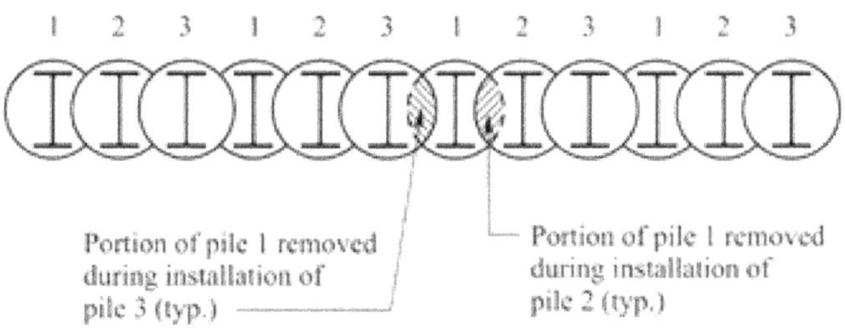

Figure: Completed Secant Pile Wall Plan View

In general, rigid support systems have more load carrying capacity than flexible systems. This additional load carrying capacity means that they require less bracing. Minimising the amount of bracing results in fewer obstruction inside the excavation if struts or braces are used, making construction activities easier to execute. Rigid wall systems incorporated into the final structure can also reduce the overall cost of the structure because they combine the support of excavation with the final structure. Waterproofing permanent support walls and detailing the connections between the walls and other structure members are difficult. This difficulty can potentially lead to leakage of groundwater into the tunnel. The design and detailing of the support of excavation must consider the sequence of installation and account for the changing loading conditions that will occur as the excavation proceeds and the system is installed.

Ground Movement and Impact on Adjoining Structures

An important issue for cut-and-cover tunnel analysis and design is the evaluation and mitigation of construction impacts on adjacent structures, facilities, and utilities. By the nature of the methods used, cut-and-cover constructions are much more disruptive than bored tunnels. It is important for engineers to be familiar with analytical aspects of evaluating soil movement as a result of the excavation, and the impacts it can have on existing buildings and utilities at the construction site. Soil movement can be due to deflection of the support of excavation walls and ground consolidation:

- *Deflection of support of excavation walls:* Walls will deflect into the excavation as it proceeds prior to installation of each level of struts or tiebacks supporting the wall. The deflection is greater for flexible support systems than for rigid systems. The deflections are not recoverable and they are cumulative.
- *Consolidation due to dewatering:* In excavations where the water table is high, it is often necessary to dewater inside the excavation to avoid instability. Dewatering inside the cut may lead to a drop in the hydrostatic pressure outside the cut. Depending on the soil strata, this can lead to consolidation and settlement of the ground.

Existing buildings and facilities must be evaluated for the soil movement estimated to occur due to the support wall movement during excavation. This evaluation depends on the type of existing structure, its distance and orientation from the excavation, the soil conditions, the type of foundations of the structure, and other parameters. The analysis is site specific, and it can be very complex. Empirical methods and screening tools are available to more generally characterise the potential impacts. The existing buildings and facilities within the zone of influence must be surveyed and monitored Geotechnical and Structural Instrumentation.

Measures to deal with this issue include:

- Design of stiffer and more watertight excavation support walls.
- Provide more closely spaced and stiffer excavation support braces and/or tiebacks.
- Use of pre-excavation soil improvement.
- Underpinning of existing structures.
- Provide monitoring and instrumentation program during excavation.
- Establish requirement for mitigation plans if movements approach allowable limits.

Base Stability

Poor soil beneath the excavation bottom may require that the excavation support structure be extended down to a more competent stratum to ensure the base stability of the structure. This may depend upon whether the earth pressures applied to the wall together with its weight can be transferred to the surrounding soil through a combination of adhesion (side friction) and end bearing.

Soft clays below the excavation are particularly susceptible to yielding causing the bottom of the excavation to heave with a potential settlement at the ground surface, or worse to blow up. High groundwater table outside of the excavation can result in base instability as well. Measures to analyse the subsurface condition, and provide sufficient base stability must be addressed by the geotechnical engineer and/or tunnel designer. Readers are referred to FHWA-NHI-05-046 "Earth Retaining Structure" (FHWA, 2005e).

Cut and Cover Tunnels

Structural Systems

General: A structural system study is often prepared to determine the most suitable structural alternatives for the construction of the cut-and-cover tunnel. This involves a determination of the proposed tunnel, the excavation support system, the tunnel structural system, the construction method (top-down vs. bottom up), and the waterproofing system. Each of these elements is interdependent upon the other. Options for each element are discussed below. The system study should consider all options that are feasible in a holistic approach, taking into account the effect that one option for an element has on another element.

Structural Element Sizing

The shape of the cut and cover tunnels is generally rectangular. The dimensions of the rectangular box must be sufficient to accommodate the clearance requirements. Dimensional information required for structural sizing includes wall heights and the span lengths of the roof. The width of the tunnel walls added to the clear space width requirements will determine the final width of the excavation required to construct the tunnel. To minimise the horizontal width of the excavation the support of excavation can be incorporated as part of, the final structure. However, this might have negative impacts on the watertightness of the structure. Some reasons that would require minimisation of the out to out width of the excavation are:

- Limited horizontal right-of-way. In urban areas where tunnels are constructed along built up city streets, additional right-of-way may be impractical to obtain. There may be existing buildings foundations adjacent to the tunnel or utilities that are impractical to move.
- There may be natural features that make a wider excavation undesirable or not feasible such as rock or bodies of water.

The depth of the roof and floor combined with the clearance requirements will define the vertical height of the tunnel structure, the depth of excavation required, and the height of the associated support of excavation. It is recommended in cut-and-cover construction that the tunnel depth be minimised to reduce the overall cost which extends beyond the cost of the tunnel structure. A shallower profile grade can also result in shorter approaches and approach grades that are more favourable to the operational characteristics of the vehicles using the tunnel resulting in lower costs for the users of the tunnel.

Structural Framing

The framing model for the tunnel will be different according to whether the support of excavation walls is a temporary (non-integral) or a permanent (integral) part of the final structure. With temporary support of excavation walls, the tunnel section would be considered a frame with fixed joints. When support of excavation walls are to form part of the tunnel structure, fixed connections between the support of excavation walls and the rest of the structure may be difficult to achieve in practice; partial fixity is more probable, but to what degree may be difficult to define. A range of fixities may need to be considered in the design analysis.

Corners of rectangular tunnels often incorporate haunches to increase the member's shear capacity near the support, in effect creating more of an arched shape. A true arch shape provides an efficient solution for the tunnel roof but tends to create other issues. Flat arches result in horizontal loads at the spring line that must be resisted by the walls. Semicircular arches eliminate these forces but result in a section larger than required vertically and drive down the tunnel profile which will add cost. When using temporary support of excavation walls, the tunnel section is constructed totally within them, often with a layer of waterproofing completely enveloping the section. In contrast, when the support of excavation walls become part of the final structure, an enveloping membrane is difficult to achieve. Therefore, provisions for overlapping, enveloping and sealing the joints would be needed. Furthermore, physical keying of the structural top and bottom slabs into the support of excavation walls is essential for any transmission of moments and shear.

Some old tunnels employ a structural system consisting of transverse structural steel frames spaced about 5 feet (1.5 m) apart. Typically, these frames are embedded in un-reinforced cast-in-place floors and walls, while for the roof, these frames are exposed and support a cast-in-place roof slab. This type of construction may still be competitive when applied to shallow tunnels, especially when longer roof spans are required for multiple lane cross sections.

Materials

Cast-in-place concrete is the most common building material used in cut and cover tunnel construction, however other materials such as precast prestressed concrete, post tensioned concrete and structural steel are used. These materials and their application are discussed below.

Cast-in-Place Concrete

Cast-in-place concrete is commonly used in tunnel construction due to the ease with which large members can be constructed in restricted work spaces. Formwork can be brought in small manageable pieces and assembled into forms for large thick members. Complex geometry can be readily constructed utilising concrete, although the formwork may be difficult to construct. Concrete is a durable material that performs well in the conditions that exist in underground structures. The low shear capacity of concrete can be offset by thickening the roof and the floor at the corners as shown.

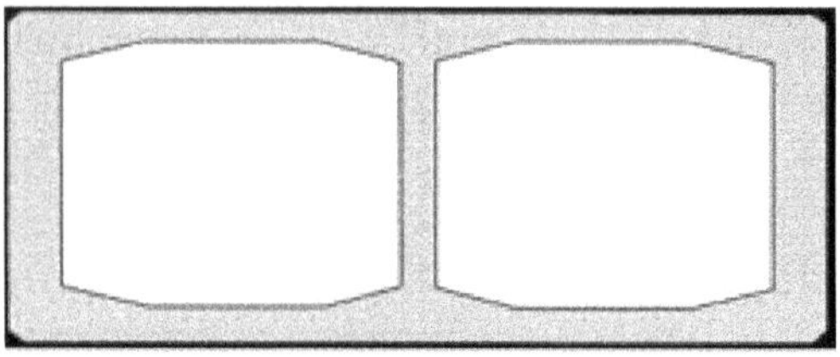

Figure: Tunnel Structure with Haunches

Connecting the structural concrete members to permanent support of excavation walls can be challenging. A simple end connection can be created quickly by placing the concrete slab in precut seats or pockets in the walls; however this results in a less efficient structure with thicker structural elements. Full moment connections can be created using splicing of reinforcing steel if sufficient wall pockets can be provided. When creating a full moment connection, the walls must be detailed to accept the transferred moment. To minimise the amount of wall pocket required, mechanical splicing or welding can be used. Waterproofing the connection, as well as the remainder of the structure, when using permanent support of excavation walls as part of the structure, is challenging.

Proper detailing of concrete members and application of all AASHTO requirements in terms of reinforcing steel is essential to create a durable concrete structure. The minimum requirements for shrinkage reinforcement should be noted. Using a larger number of smaller bars rather than a small number of large bars helps distribute cracks and consequently reduces their size. Ground water chemistry should be investigated to ensure that proper mix designs compatible with ground water chemistry are used to reduce the potential for chemical attack of the concrete.

Structural Steel

Structural steel has excellent weight to strength characteristics. Structural steel beams with a composite slab can be used to reduce the thickness of roof slabs. This can reduce the depth to the profile with the accompanying reductions in overall cost of the tunnel associated with a shallower excavation and shorter retained tunnel approaches. Structural steel is easier to connect to permanent support of excavation walls than are concrete slabs. Local removal of the permanent wall in small isolated pockets is all that is required to provide a seat for the steel beam creating a simple end. If simple ends are used, the movement of the beam due to temperature changes inside the tunnel should be accommodated. If the support of excavation used SPTC, tangent or secant pile walls, the embedded steel cores of these walls can be exposed and a full moment connection can be made. A full moment connection will not allow temperature movements, so the resulting force effects must be evaluated and accommodated by design. Structural steel beams are best fabricated and delivered in a single piece. However, if the excavation support system has complex internal bracing, it may not be possible to deliver and erect the steel beams inside the excavation which would require splicing of the steel beams. Connections also require careful inspection which adds to the future maintenance cost of the tunnel if the connections are not encased. Waterproofing the connections to the exterior walls can be difficult. Tunnels typically produce a damp environment, if combined with the potential to leak around connections, this results in conditions that can result in aggressive corrosion to steel members. Corrosion protection must be considered as part of the structural steel structural system.

In addition to the roof structure described above, steel frames have also been used in road tunnels and under some circumstances, may still be appropriate. The frame includes columns and the roof beams. In permanent support walls, the columns would be embedded in the walls. The steel columns are erected on a suitable foundation cast on the bottom of the excavation, the beams are then erected and joined with the columns and the entire frames are then encased in concrete, with nominal reinforcement. The roof beams can be completely encased or exposed supporting a thin concrete roof slab. If exposed, inspection and maintenance are required.

Prestressed Concrete

Prestressed concrete, including precast prestressed beams such as AASHTO beams or similar, may be suitable for large roof spans when clearances are tight and the overall depth of section must be limited. Precast prestressed beams have been used for the top slab supported on cast in place walls. Precast concrete beams, in the number and lengths required for cut and cover tunnels are impractical to splice. They must be delivered in a single piece and be able to be erected within the space available inside the excavation. The type and configuration of the excavation must therefore be considered when evaluating the use of precast concrete beams. Making connections with permanent support of excavation walls can be accomplished by creating pockets in the walls to support the beams in a simple support arrangement. Simple supports also require a method for allowing movement of the beams during temperature changes inside the tunnel. Waterproofing this connection is difficult. Making a moment connection requires more elaborate details of the junction between the wall and the beam to be able to install the reinforcing required for the moment connection. A moment connection at the beam also requires that the wall itself be capable of accepting the moment transferred by the beam. Therefore the detailing of the wall must be compatible with the structural system selected. A full moment connection will not allow temperature movements, so the resulting force effects must be evaluated and accommodated by the design.

Although seldom, post tensioning is used in cut and cover tunnels; however in developing the post tensioning strategy, it is important to consider the various loading stages and potentially have multiple stages of post tensioning. For example, the introduction of high post-tensioning forces in tunnel slabs before backfilling causes temporary high tensile stresses in the opposite face of the slabs. These stresses may limit the depth to which post-tensioned members can be used, unless some of the tendons are tensioned from inside the box after backfilling. The elastic shortening of the slab will induce resistance to the post-tensioning via the walls, and should be taken into consideration. The additional moments created will also need to be resisted. Isolating the top slab from the walls by means of a movement joint (such as neoprene or Teflon bearings) would eliminate the above shortcomings but also eliminate the advantages of moment connection; waterproofing of the movement joint will need to be addressed. The design should identify space requirements for operation of the stressing jacks from both sides (if required). In many cases, the tendon would be less than 100 ft (30 m) long, needing only one end for stressing. Usually, in such a case, alternate strands would be stressed from alternate ends, requiring suitable space on each side.

Buoyancy

Buoyancy is a major concern in shallow tunnels that are under or partially within the water table. Buoyancy should be checked during the design. The structural system selected should take into account its ability to resist buoyancy forces with its own weight or by providing measures to deal with negative buoyancy. In cases where the structure and backfill are not heavy enough to resist the buoyancy forces, flotation can occur. Measures to resist the forces of flotation must be provided and accounted for in the design.

The resistance against flotation can be achieved by a variety of methods. Typical methods used to increase the effective weight of the structure include:

- Connecting the structure to the excavation support system and thus mobilising its weight and/or its friction with the ground
- Thickening structural members beyond what is required for strength in order to provide dead load to counter the flotation forces
- Widening the floor slab of the tunnel beyond the required footprint to key it into adjacent soil and thus to include the weight of soil above these protrusions
- Using steel or concrete tension piles to resist the uplift forces associated with flotation
- Using permanent tie-down anchors; in soils, it may be prudent for the anchors only to carry a nominal tension under normal conditions and for the anchors to be fully mobilised only under extreme conditions. Properly protected anchor heads can be located in formed recesses within the base slab
- Permanent pressure relief system beneath the base of the structure. This is a complicated system to remove the buoyant forces by allowing water to be collected from under the bottom slab and removed from the tunnel. This type of system requires maintenance and redundancy in addition to the life cycle costs associated with operating the system. It can also have the effect of lowering the local groundwater table which may have negative consequences.

Considering the long design life of underground structures, the design of tension piles or tie-down anchors to resist flotation forces must include provisions to address the risk of corrosion of these tension elements and consideration of their connection to the tunnel structure. Similarly, the use of an invert pressure relief system and backup system must include provisions to address the risk of the long-term operation and maintenance requirements. For most projects, generally, buoyancy forces are resisted by increased dead load of the structure and/or weight of fill above the structure.

Expansion and Contraction Joints

Many cut and cover tunnels are constructed without permanent expansion or contraction joints. Although expansion joints may not be required except close to the portals, contraction joints are recommended throughout the tunnel.

Significant changes in support stiffness or surcharge can cause differential settlement. If the induced moments and shears resulting from this are greater than the section can handle, relieving joints can be used to accommodate localised problems. Expansion joints are usually provided at the interfacing with ventilation building or portals or other rigid structures to allow for differential settlements and movements associated with temperature changes. It is recommended that contraction joints be placed at intervals of approximately 30 feet (about 9 m).

Seismic loading can cause significant bending moments in cut and cover tunnels. Joints may be used to relieve the moments and shears that would have occurred in continuous rigid structures, particularly as the width (and hence the stiffness) of the structure increases. Joints may also be required to handle relative seismic motion at locations where the cross-sectional properties change significantly, such as at ventilation buildings and portals. Such motion can be both longitudinal and transverse (horizontal and vertical) to the tunnel.

Joints are potential areas were leaks can occur. As such, they are potential sources of high maintenance costs over the life of the tunnel. The number of joints should be minimised and special care should be taken in the detailing of joints to ensure water tightness. The type and frequency of joints required will be a function of the structural system required and should be evaluated in the overall decision of the type selected.

Waterproofing

The existence of a high groundwater table or water percolating down from above requires that tunnels be waterproof. Durability is improved when the tunnel is waterproof. Good waterproofing design is also imperative to keep the tunnel dry and reduce future maintenance. Leaking tunnels are unsightly and can give rise to concern by users. In colder climates such as in the North East, leaks can become hazardous ceiling icicles or ice patches on roadways. The waterproofing system should be selected based on the required performance and its compatibility with the structural system.

Loads

General: The relevant loads to be considered in the design of the cut and cover tunnel structures along with how to combine the loads of the AASHTO LRFD specifications. The AASHTO LRFD specification divides loads into two categories: Permanent Loads and Transient Loads. "Load and Load Designation" of the AASHTO LRFD specifications defines following permanent loads that are applicable to the design of cut and cover tunnels:

DC = Dead Load: This load comprises the self weight of the structural components as well as the loads associated with nonstructural attachments. Nonstructural attachments can be signs, lighting fixtures, signals, architectural finishes, waterproofing, etc. Typical unit weights for common building materials are given in Table 3.5.1-1 of the AASHTO LRFD specifications. Actual weights for other items should be calculated based on their composition and configuration.

DW = Dead Load: This load comprises the self weight of wearing surfaces and utilities. Utilities in tunnels can include power lines, drainage pipes, communication lines, water supply lines, etc. Wearing surfaces can be asphalt or concrete. Dead loads, wearing surfaces and utilities should calculate based on the actual size and configuration of these items.

EH = Horizontal Earth Pressure Load. The information required to calculate this load is derived by the geotechnical data developed during the subsurface investigation program. In lieu of actual subsurface data, the information contained in paragraph of the AASHTO specifications can be used. At-rest pressures should be used in the design of cut and cover tunnel structure.

EL = Accumulated locked-in force effects resulting from the construction process including secondary forces from post tensioning if used.

ES = Earth surcharge load. This is the vertical earth load due to fill over the structure that was placed above the original ground line. It is recommended that a minimum surcharge load of 400 psf be used in the design of cut and cover tunnels. *If there is a potential for future development over the tunnel structure, the surcharge from the actual development should be used in the design of the structure. In lieu of a well defined loading, it is recommended that* a minimum value of 1000 psf be used when future development is anticipated.

EV = Vertical pressure from the dead load of the earth fill. This is the vertical earth load due to fill over the structure up to the original ground line. The information required to calculate this load are derived by the geotechnical data developed during the subsurface investigation program. In lieu of actual subsurface data, the information contained of the AASHTO specifications can be used.

The LRFD specifications defines following transient loads that are applicable to the design of cut and cover structures:

CR = Creep

CT = Vehicular Collision Force: This load would be applied to individual components of the tunnel structure that could be damaged by vehicular collision. Typically, tunnel walls are very massive or are protected by redirecting barriers so that this load need be considered only under usual circumstances. It is preferable to detail tunnel structural components so that they are not subject to damage from vehicular impact.

EQ = Earthquake. This load should be applied to the tunnel lining as appropriate for the seismic zone for the tunnel. The scope of this manual does not include the calculation of or design for seismic loads. However, some recommendations are provided. The designer should be aware that seismic loads should be accounted for in the design of the tunnel lining in accordance with LRFD Specifications.

IM = Vehicle Dynamic Load Allowance: This load can apply to the roadway slabs of tunnels and can also be applied to roof slab of tunnels that are constructed under other roadways, rail lines, runways or other facilities that carry moving vehicles.

LL = Vehicular Live Load: This load can apply to the roadway slabs of tunnels and can also be applied to roof slab of tunnels that are constructed under other roadways, rail lines, runways or other facilities that carry moving vehicles. This load would be distributed through the earth fill prior to being applied to the tunnel roof, unless traffic bears directly on the tunnel roof. Guidance for the distribution of live loads to buried structures can be found.

SH = Shrinkage: Cut and cover tunnel structural elements usually are relatively massive. As such, shrinkage can be a problem especially if the exterior surfaces are restrained. This load should be accounted for in the design or the structure should be detailed to minimise or eliminate it.

TG = Temperature Gradient: Cut and cover structural elements are typically constructed of concrete which has a large thermal lag. Combined with being surrounded by an insulating soil backfill that maintains a relatively constant temperature, the temperature gradient across the thickness of the members can be measurable. This load should be examined on case by case basis depending on the local climate and seasonal variations in average temperatures. The AASHTO LRFD specifications provides guidance on calculating this load.

TU = Uniform Temperature: This load is used primarily to size expansion joints in the structure. If movement is permitted at the expansion joints, no additional loading need be applied to the structure. Since the structure is rigid in the primary direction of thermal movement, the effects of the friction force resulting from thermal movement can be neglected in the design. Some components may be individually subject to this load. The case where concrete or steel beams support the roof slab is an example. If these beams are framed into the side walls to create a full moment connection, the expansion and contraction of these beams will add force effect to the frames formed by the connection. This effect must be accommodated in the design. This effect is usually not considered in the case of a cast-in-place concrete box structure due to the insulating qualities of the surrounding ground and the large thermal lag of concrete.

WA = Water Load: This load represents the hydrostatic pressure expected outside the tunnel structure. Tunnel structures are typically detailed to be watertight without provisions for relieving the hydrostatic pressure. As such, the tunnel is subject to horizontal hydrostatic pressure on the sidewalls, vertical hydrostatic pressure on the roof and a buoyancy force on the floor. Hydrostatic pressure acts normal to the surface of the tunnel. It should be assumed that water will develop full hydrostatic pressure on the tunnel walls, roof and floor. The design should take into account the specific gravity of the groundwater which can be saline near salt water. Both maximum and minimum hydrostatic loads should be used for structural calculations as appropriate to the member being designed. For the purpose of design, the hydrostatic pressures assumed to be applied to underground structures should ignore pore pressure relief obtained by any seepage into the structures unless an appropriately designed pressure relief system is installed and maintained. Two groundwater levels should be considered: normal (observed maximum groundwater level) and extreme, 3 ft (1 m) above the design flood level (100 to 200 year flood).

DD = Downdrag: This load comprises the vertical force applied to the exterior walls of a top-down structure that can result from the subsidence of the surrounding soil due to the subsidence of the in-situ soil below the bottom of the tunnel. This load would not apply to cut and cover structures since it requires subsidence or settlement of the material below the bottom of the structure to engage the downdrag force of the walls. For the typical highway tunnel, the overall weight of the structure is usually less than the soil it is replacing. As such, unless backfill in excess of the original ground elevation is paced over the tunnel or a structure is constructed over the tunnel, settlement will not be an issue for cut and cover tunnels.

BR = Vehicular Breaking Force: This load would be applied only under special conditions where the detailing of the structure requires consideration of this load. Under typical designs, this force is resisted by the mass of the roadway slab and need not be considered in design.

CE = Vehicular Centrifugal Force: This load would be applied only under special conditions where the detailing of the structure requires consideration of this load. Under typical designs, this force is resisted by the mass of the roadway slab and need not be considered in design.

CV = Vessel Collision Force is generally not applicable to cut and cover construction unless it is done under a body of water such as in a cofferdam. It is applicable to immersed tube tunnels, which are a specialised form of cut and cover tunnel and are covered separately.

FR = Friction: As stated above, the structure is usually rigid in the direction of thermal movement. Thermal movement is the source of the friction force. In a typical tunnel, the effects of friction can be neglected.

IC = Ice load: Since the tunnel is not subjected to stream flow nor exposed to the weather in a manner that could result in an accumulation of ice, this load is not used in cut and cover tunnel design.

PL = Pedestrian Live Load: Pedestrian are typically not allowed in road tunnels, so there is no need to design for a pedestrian loading.

SE = Settlement: For the typical road tunnel, the overall weight of the structure is usually less than the soil it is replacing. As such, unless backfill in excess of the original ground elevation is paced over the tunnel or a structure is constructed over the tunnel, settlement will not be an issue for cut and cover tunnels. If settlement is anticipated due to poor subsurface conditions or due to the addition of load onto the structure or changing ground conditions along the length of the tunnel, it is recommended that ground improvement measures or deep foundation (piles or drilled shafts) be used to support the structure.

WL = Wind on Live Load: The tunnel structure is not exposed to the environment, so it will not be subjected to wind loads.

WS = Wind Load on Structure: The tunnel structure is not exposed to the environment, so it will not be subjected to wind loads.

The LRFD specifications provides guidance on the methods to be used in the computations of these loads. The design example in Appendix C shows the calculations involved in computing these loads. The order of construction will impact loading and assumptions. For example, in top down construction, permanent support of excavation walls used as part of the final structure will receive heavier bearing loads, because the roof is placed and loaded before the base slab is constructed. The permanent support of excavation walls are also braced as the excavation progresses by the roof slab resulting a different lateral soil pressure distribution than would be found in the free standing walls of a cast-in-place concrete structure constructed using bottom up construction. The base slab of a top-down construction tunnel acts as a mat for supporting vertical loads, but it is not available until towards the end of construction of the section eliminating its use to resist moments from the walls or to act as bracing for the walls. Typical loading diagrams are illustrated respectively for bottom-up and top-down structures.

Structural Analysis

Structural analysis is covered in Section 4 of the AASHTO LRFD specifications. It is recommended that classical force and displacement methods be used in the structural analysis of cut and cover tunnel structures. Other numerical methods may be used, but will rarely yield results that vary significantly from those obtained with the classical methods. The modelling should be based on elastic behaviour of the structure as per the AASHTO LRFD specifications.

Since all members of a cut and cover tunnel, with the possible exception of the floor of tunnels built using top-down construction, are subjected to bending and axial load, the secondary affects of deflections on the load affects to the structural members should be accounted for in the analysis. The AASHTO LRFD specifications refer to this type of analysis as "large deflection theory". Most general purpose structural analysis software have provisions for including this behaviour in the analysis. If this behaviour is accounted for in the analysis, no further moment magnification is required.

The AASHTO LRFD specifications states that the design of the structure should include "...where appropriate, response characteristics of the foundation". The response of the foundation for a cut and cover tunnel structure can be modelled through the use of a series of non linear springs placed along the length of the bottom slab. These springs are non linear because they should be specified to act in only one direction, the downward vertical direction. This model will provide the proper distribution of loads to the bottom slab of the model and give the designer an indication if buoyancy is a problem. This indication is seen in observing the calculated displacements of the structure. A net upward displacement of the entire structure indicates that there is insufficient resistance to buoyancy.

Structural models for computer analysis are developed using the centroid of the structural members. As such, it is important when calculating the applied loads, that the loads are calculated at the outside surface of the members. The load is then adjusted according to the actual length of the member as input.

Other numerical methods of analysis for cut-and-cover tunnel sections include:

- Frame analysis with a more rigorous soil-structure interaction by modelling the soil properties together with the tunnel. The same frame analysis, but with the addition of a series of unidirectional springs on the underside to model the effect of the soil as a beam on an elastic foundation. Lateral or horizontal springs may be applied in conjunction with assumed soil loads. Care must be taken to ensure that the assumed soil spring acts only when deflection into the soil occurs. This may require multiple iterations of the input parameters for each load combination. Many commercially available programs will automatically adjust the input values and rerun the analysis. This gives a better modelling representation of the structure and takes advantage of more realistic base slab soil support, often resulting in more economical design. Setting up a model is a little more difficult with the springs, and suitable values for the spring modulus are difficult to quantify. It may be appropriate to use a range of values and run the model for each.
- Finite element and finite difference analyses. The material of the tunnel structure and the soil are modelled as a continuum grid of geometric elements. Structural elements are usually treated as linear elastic. A number of different mathematical models for the soil type are available. This method of modelling and analysis can more closely represent actual conditions, especially if better numerical resolution is used where there

are conditions of difficult tunnel geometry such as the framing details. The method is usually complex to setup and run, and results require careful interpretation.

As stated above, two-dimensional sectional analysis is sufficient for most tunnel conditions. Three-dimensional modelling may be required where tunnel sections vary along the length of the tunnel or where intersections exist such as at ramps or cross-passages. 3-D modelling is very complex and the accuracy of the loading data, uncertainty about soil behaviour, and its inherent lack of homogeneity may not warrant such detailed analysis for highway tunnels except for special locations such as ramps, cross passages, and connections to other structures.

Groundwater Control

Construction Dewatering

When groundwater levels are higher than the base level of the tunnel, excavations will require a dewatering system. For cut and cover construction, the dewatering systems will depend on the permeability of the various soil layers exposed. Lowering the water table outside the excavation could cause settlement of adjacent structures, impact on vegetations, drying of existing wells, and potential movement of contaminated plumes if present. Precautions should be taken when dewatering the area outside the excavation limits. Within the excavation, dewatering can be accomplished with impermeable excavation support walls that extend down to a firm, reasonably impermeable stratum to reduce or cut-off water flow.

Impervious retaining walls, such as steel interlocking sheeting or concrete slurry walls, could be placed into deeper less pervious layers, such as glacial till or clay, to reduce groundwater inflow during construction and limit draw-down of the existing groundwater table. For most braced excavation sites, dewatering within the excavation is often done. Sometimes the excavation is done in the wet, then the water is pumped out. Subsequent to the excavation, any water intrusion will be pumped from the trench by providing sumps and pumps within the excavation. In some areas, a pumped pressure relief system may be required to prevent the excavation bottom from heaving due to unbalanced hydrostatic pressure.

Pumped wells can be used to temporarily lower the groundwater table outside the excavation support during construction; however this may have environmental impact or adverse effects on adjacent structures. To minimise any lowering of the water table immediately outside the excavation, water pumped from the excavation can be used to recharge the water bearing strata of the groundwater system by using injection wells. Provision would have to be made for disposal of water in excess of that pumped to recharge wells, probably through settlement basins draining to storm drains.

After construction is completed, if there is a concerned that the permanent excavation support walls above the tunnel might be blocking the cross flow of the groundwater or may dam up water between walls above the tunnel, the designer may need to consider to breach the walls above the tunnel at intervals or removed to an elevation to allow movement of groundwater. Granular backfill around tunnels can also help to maintain equal hydrostatic heads across underground structures.

Methods of Dewatering and Their Typical Applications

Groundwater can be controlled during construction either by using impervious retaining walls (such as concrete slurry or tangent pile walls, steel interlocking sheeting, etc.), by well-points drawing down the water table, by chemical or grout injection into the soils, or by pumping from within the excavation. Groundwater may be lowered, as needed, by tiers of well-points. Improper control of groundwater is often a cause for settlement and damage to adjacent structures and utilities; consequently it is important that the method selected is suitable for the proposed excavation. Where the area of excavations is not too large, an economical method of collecting water is through the use of ditches leading to sump pumps. Provisions to keep fines from escaping into the dewatering system should be made.

In larger excavations in permeable soil, either well points or deep wells are often used to lower the water table in sand or coarse silt deposits, but are not useful in fine silt or clay soils due to their low permeability. It is recommended that test wells be installed to test proposed systems. In certain cases, multiple stages of well points, deep wells with submersible pumps or an eductor system would be needed.

Uplift Pressures and Mitigation Measures

Piping and Base Stability: In fine-grained soils, such as silts or clayey silts, differential pressure across the support of excavation may cause sufficient water flow (piping) for it to carry fines. This causes material loss and settlement outside as well as a loss of integrity of soils within, rendering the soils unsuitable as a foundation. In extreme cases, the base of the excavation may become unstable, causing a blow-up and failure of the excavation support. This situation may be mitigated by ensuring that cut-off walls are sufficiently deep, by stabilising the soil by grouting, or freezing, or by excavating below water without dewatering and making a sufficiently thick tremie slab to overcome uplift before dewatering.

Potential Impact of Area Dewatering

Dewatering an excavation may lower the groundwater outside the excavation and may cause settlements. The lowering of the external groundwater can be reduced by the use of slurry walls, tangent or secant piles, or steel sheet piling. Adjacent structures with a risk of settlement due to groundwater lowering may require underpinning. Furthermore, where lowering of groundwater exposes wooden piles to air, deterioration may occur.

Groundwater Discharge and Environmental Issues

In most cases, the water will require testing and possibly treatment before it can be discharged. Settling basins, oil separators, and chemical treatments may be required prior to disposal. Local regulations and permitting requirements often dictate the method of disposal.

The excavated material itself will require testing before the method of disposal can be determined. Material excavated below water may need to stand in settling ponds to allow excess water to run off before disposal. Contaminated material may need to be placed in confined disposal facilities.

Maintenance and Protection of Traffic

When the excavation crosses existing roads or is being performed under an existing road, decking would be required to maintain the existing road traffic. When decking is required the support of excavation walls must be designed to handle the imposed live loads. The depth of the walls may need to be determined by the necessity of transferring decking loads to a more competent stratum below. This may depend upon whether the load applied to the wall together with its weight can be transferred to the surrounding soil through a combination of adhesion (side friction) and end bearing. Thick types of excavation support walls, such as slurry walls, drilled-in-place soldier piles, and tangent piles, are much more effective than thinner walls, such as sheet piles or driven soldier piles, in carrying the live loads to the bearing stratum. Decking often consists of deck framing and roadway decking. A typical general arrangement for street decking over a cut-and-cover excavation using timber decking. Pre-cast concrete planks have been used also as decking. Structural steel deck beams can be arranged to function also as the uppermost bracing tier of the support of excavation. The deck framing should be designed for AASHTO HL-

93 loading, or for loading due to construction equipment that actually will operate on the deck, whichever is greater.

Utility Relocation and Support

Types of Utilities: Constructing cut and cover tunnels in urban areas often encounters public and private utility lines such as water, sewer, power, communication, etc... Often, utilities are not located where indicated on existing utility information.

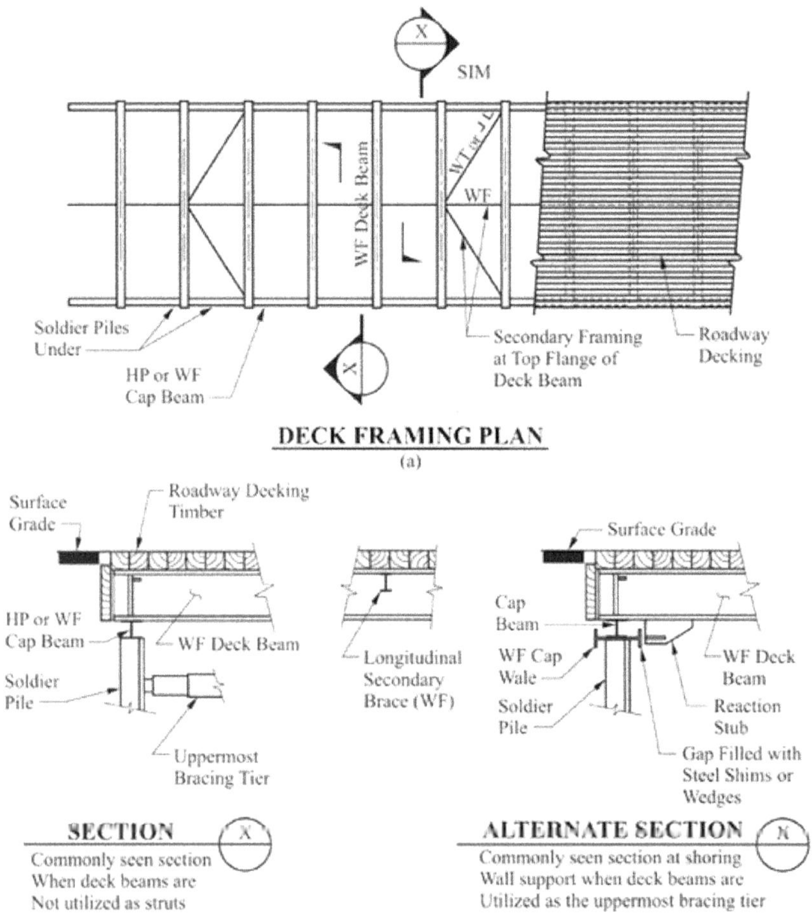

Figure: Typical Street Decking (After Bickel et al., 1996)

Therefore, it is important to identify and locate all utilities in the field prior to excavation. Great care must be taken when excavating in the vicinity of utilities, sometimes requiring that the final excavation to expose them be done by hand. Of particular concern are those utilities that are movement-sensitive and those carrying hazardous substances; these include large diametre water pipes, high pressure gas lines, fibre optic lines, petroleum pipes and high voltage cables. Some utilities such as buried high voltage lines are not only extremely expensive to move but have very long lead times. Utilities such as sewers can present a different problem; if gravity flow is used, diversions around a proposed tunnel may pose a serious challenge. Some older water and sewer lines are extremely fragile, particularly if they are of brick or cast iron construction.

General Approach to Utilities During Construction

It is not uncommon to divert utilities away from the proposed construction corridor. However, diversion is not always possible, it may be too expensive, or a utility crossing may be unavoidable; in such cases, it will be necessary to support the utilities in place. It is essential to have a coordinated effort so that no interferences among the various utilities occurs and that the construction can be done while the utilities are in place. Sometimes, utility relocations are done in stages to accommodate the construction requiring relocating the utility more than once. Before the start of underground construction, a condition survey should be made of all utilities within the zone of potential influence of construction, making detailed reports for those that may incur movements in excess of those allowable for the utility. The nature of any work required for each utility should be identified, i.e., protection, support or relocation, and the date by which action is required. It is essential that all utilities that need action are identified in sufficient time to allow the construction to progress as programmed. Supports may either be temporary or permanent. Depending upon the sensitivity of the utility being supported, it may be necessary to provide instrumentation to monitor any movement so that remedial action can be taken before damage occurs. Systems providing vertical support should be designed as bridge structures. Lateral support may be considered as retaining walls.

Most utilities require access for repairs; it is therefore required to have provisions for access to utilities passing beneath a tunnel. In some cases, it has been found appropriate to relocate utilities to a trough or utility tunnel in which all utilities can be easily accessed. In some cases, utilities cannot be raised sufficiently to clear the tunnel roof slab; it may be possible to create a narrow trough across the roof in which the utility may be relocated. In certain situations, utilities were passed through the tunnel by providing a special conduit below the tunnel roof. In all cases, all utility work must be carefully coordinated with the utility owner.

Chapter 4

Rock Tunnelling

Design recommendations and requirements for mined and bored road tunnels in all types of grounds. Design and construction issues for rock tunnelling including rock failure mechanism, rock mass classification, excavation methods, excavation supports, and design considerations for permanent lining, groundwater control, and other ground control measures. Because of the range of behaviour of tunnels in rock, i.e., from a coherent continuum to a discontinum, stabilisation measures range from no support to bolts to steel sets to heavily reinforced concrete lining and numerous variations and combinations in between. Certainly these variations are to be expected when going from one tunnel to another but often several are required in a single tunnel because the geology and/or the geometry change.

Thus, the engineer must recognise the need for change and prepare the design to allow for adjustments to be made in the field to adjust construction means, methods, and equipment to the challenges presented by the vagaries of nature. This chapter provides the engineer with the basic tools to approach the design, it is not a cookbook that attempts to give instantaneous solutions/designs for the novice designer. The data needed for analysis and design rock tunnels and the investigative techniques to obtain the data. The results of the analysis and design presented hereafter are typically presented in the geotechnical/technical design memorandum and form the basis of the Geotechnical Baseline

Rock Failure Mechanism

Only in the last half-century has rock mechanics evolved into a discipline of its own rather than being a sub-set of soil mechanics. At the same time there was a "merging of elastic theory, which dominated the English language literature on the subject, with the discontinuum approach of the Europeans" (Hoek, 2000). These two phenomena have also occurred during a time of ever-increasing demand for economical tunnels. Hence, design and construction of rock tunnels have taken on a new impetus and importance in the overall field of heavy construction as it applies to infrastructure.

Understanding the failure mechanism of a rock mass surrounding an underground opening is essential in the design of support systems for the openings. The failure mechanism depends on the in situ stress level and characteristics of the given rock mass. At shallow depths, where the rock mass is blocky and jointed, the stability problems are generally associated with gravity falls of wedges from the roof and sidewalls since the rock confinement is generally low. As the depth below the ground surface increases, the rock stress increases and may reach a level at which the failure of the rock mass is induced. This rock mass failure can include spalling, slabbing, and major rock burst.

Conversely, excavation of an underground opening in an unweathered massive rock mass may be the most ideal condition. When this condition, paired together with relatively low stresses, exists, the excavation will usually not suffer from serious stability problems, thus support requirement will be minimal.

Wedge Failure

Due to the size of tunnel openings (relative to the rock joint spacing) in most infrastructure applications, the rock around the tunnel tends to act more like a discontinuum. Behaviour of a tunnel in a continuous material depends on the intrinsic strength and deformation properties of that material whereas behaviour of a tunnel in a discontinuous material depends on the character and spacing of the discontinuities. Design of the former lends itself more naturally to analytical modelling (similar to most tunnels in soil) whereas design of the latter requires consideration of possible block or wedge movement or failure wherein the design approach is to hold the rock mass together. By doing so, the rock is forced to form a "ground arch" around the opening and hence to redistribute the forces such that the ground itself carries most of the "load".

To stabilise blocks or wedges, and hence the opening, the first step is to determine the number, orientations and conditions of the joints. The Q system, described in 6.3.4 gives the basic information required for the joint sets:

- Number of joints
- Joint roughness

- Joint alteration
- Joint water condition
- Joint stress condition

With these parameters defined, analyses can be made of the block or wedge stability and of the support required to increase that stability to a satisfactory level. For small tunnels of ordinary geometry the initial analysis (if not the final) can be estimated from a simple free-body approach. For larger tunnels with complicated geometry and/or a more complicated joint system, it is recommended that a computer program such as Unwedge be used to analyse the opening. Once the basic parameters of the problem are input to the program, a series of runs can be made to evaluate the impact of such variations on the calculated support required for the opening.

As indicated earlier, except for a small tunnel in very massive rock, the concept of "solid rock" is usually a misconception. As a result, the behaviour of the ground around a rock tunnel is usually the combination of that of a blocky medium and a continuum. Hence, the "loads" on the tunnel support system are usually erratic and nonuniform. This is in contrast to soft ground tunnels where the ground may sometimes be approximated by elastic or elastic-plastic assumptions or where the parameters going into numerical modelling are significantly more amenable to rational approximations.

In its simplest terms the challenge to supporting a tunnel in rock is to prevent the natural tendency of the rock to "unravel". Most failures in rock tunnels are initiated by a block (called "keyblocks" by Goodman, 1980) that wants to loosen and come out. When that block succeeds, others tend to loosen and follow. This can continue until the tunnel completely collapses or until the geometry and stress conditions come to equilibrium and the unravelling stops. Contrarily, if that first block can be held in place the stresses rearrange themselves into the ground arch around the tunnel and stability is attained.

Stress Induced Failure

As the depth of a tunnel becomes greater or where adjacent underground structures exist and the ground condition becomes less favourable, the stress within the surrounding rock mass increases and failure occurs when the stress exceeds the strength of the rock mass. This failure can range from minor spalling or slabbing in the rock surface to an explosive rockburst where failure of a significant volume of rock mass occurs.

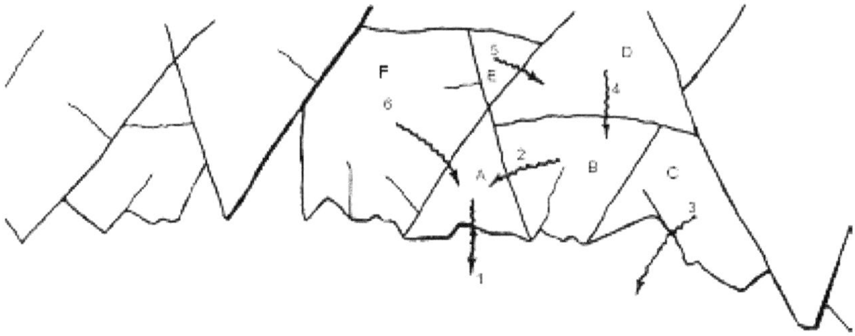

- Step 1- Block A drops down
- Step 2- Block B rotates counterclockwise and drops out
- Step 3- Block C rotates counterclockwise and drops out
- Step 4- Block D drops out followed by block E
- Step 5- Block E drops out followed by block F
- Step 6- Block F rotates clockwise and drops out

Figure: Progressive Failure in Unsupported Blocky Rock

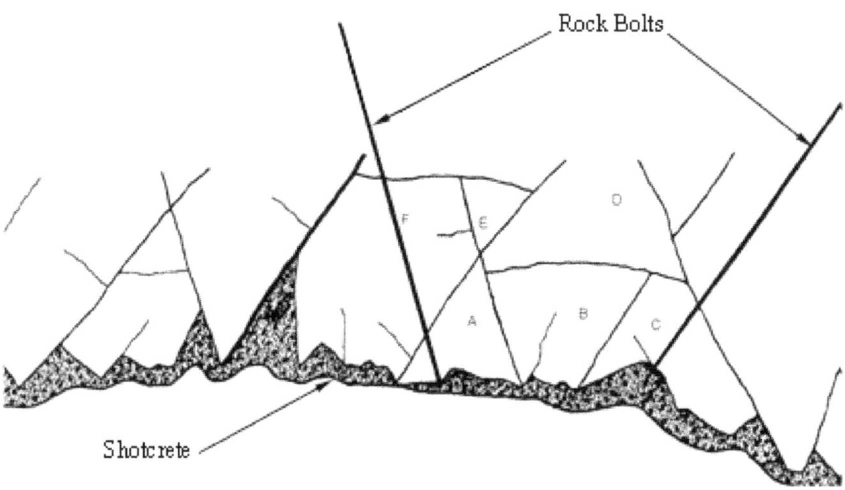

- Step 1- Block A and C are held in place by rock bolts and shotcrete
- Step 2- Block B is held in place by Blocks A and C
- Step 3- Block D is held in place by Blocks A, B, and C
- Step 4- Blocks E and F are held in place by Blocks A, B, and D assisted by rock bolts and shotcrete

Figure: Prevention of Progressive Failure in Supported Blocky Rock

The stress induced failure potential can be investigated using the strength factor (SF) against shear failure defined as $(\sigma_{1f} - \sigma_3)/(\sigma_1 - \sigma_3)$, where $(\sigma_{1f} - \sigma_3)$ is the strength of the rock mass and $(\sigma_1 - \sigma_3)$ is the induced stress, σ_1 and σ_3 are major and minor principal stresses, and σ_{1f} is major principal stress at failure. A SF greater than 1.0 indicates that the rock mass strength is greater than the induced stress, i.e., there is no overstress in the rock mass. When SF is less than 1.0, the induced stresses are greater than the rock mass strength, and the rock mass is overstressed and likely to behave in the plastic range.

Squeezing and Swelling

Squeezing rock is associated with the creation of a plastic region around an opening and severe face instability. From a tunnel design point of view, a rock mass is considered to be weak when its in-situ uniaxial compressive strength is significantly lower than the natural and excavation induced stresses acting upon the rock mass surrounding a tunnel. Hoek et. al. (2000) proposed a chart to predict squeezing problems based on strains with no support system as shown. As a very approximate and simple estimation, figure can be directly used to predict squeezing potential by comparing rock mass strength and in-situ stress. If finite element analysis results are available, one can simply predict the squeezing potential based on the calculated strains from the FE analysis. For example, the squeezing problems, if a tunnel is excavated at the proposed depth, are severe when the calculated strains from FE analysis is 2.5% or higher.

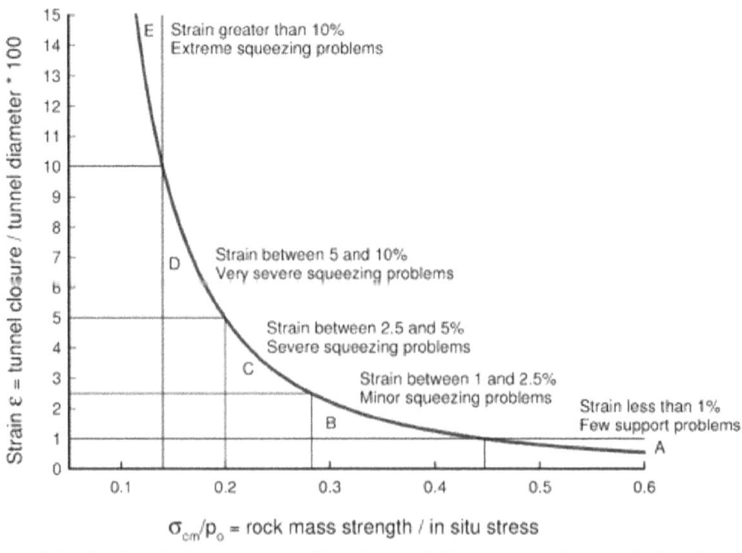

Figure: A Relationship between Strain and Squeezing Potential of Rock Mass (Hoek, et. al., 2000)

Swelling rock, in comparison, is associated with an increase in moisture content of the rock. Swelling rock can sometimes be associated with squeezing rock, but may occur without formation of a plastic zone. The swelling is usually associated with clay minerals, indurated to shale or slate or not, imbibing water and expanding. A relatively simple swell test in the laboratory will allow prediction of the swell and will also provide the "swelling pressure", where the swelling pressure is defined as that pressure that must be applied to the rock to arrest the swelling. Obviously, the support system has to resist at least the full swelling pressure to arrest the swelling movement. Montmorillinitic shales, weathered nontronite basalts, and some salts found in evaporate deposits are typical swelling rocks.

Rock Mass Classifications

Rock mass classification schemes have been developed to assist in (primarily) the collection of rock into common or similar groups. The first truly organised system was proposed by Dr. Karl Terzaghi (1946) and has been followed by a number of schemes proposed by others. Terzaghi's system was mainly qualitative and others are more quantitative in nature. The following subsections explain three systems and show how they can be used to begin to develop and apply numerical ratings to the selection of rock tunnel support and lining. This section discusses various rock mass classification systems mainly used for rock tunnel design and construction projects.

Terzaghi's Classification

Today rock tunnels are usually designed considering the interaction between rock and ground, i.e., the redistribution of stresses into the rock by forming the rock arch. However, the concept of loads still exists and may be applied early in a design to "get a handle" on the support requirement. The concept is to provide support for a height of rock (rock load) that tends to drop out of the roof of the tunnel (Terzaghi, 1946). Terzaghi's qualitative descriptions of rock classes are summarised in Table 6-1.

Table 1: Terzaghi's Rock Mass Classification

Rock Condition	*Descriptions*
Intact rock	Contains neither joints nor hair cracks. Hence, if it breaks, it breaks across sound rock. On account of the injury to the rock due to blasting, spalls may drop off the roof several hours or days after blasting. This is known

	as a spalling condition. Hard, intact rock may also be encountered in the popping condition involving the spontaneous and violent detachment of rock slabs from the sides or roof
Stratified rock	Consists of individual strata with little or no resistance against separation along the boundaries between the strata. The strata may or may not be weakened by transverse joints. In such rock the spalling condition is quite common
Moderately jointed rock	Contains joints and hair cracks, but the blocks between joints are locally grown together or so intimately interlocked that vertical walls do not require lateral support. In rocks of this type, both spalling and popping conditions may be encountered
Blocky and seamy rock	Consists of chemically intact or almost intact rock fragments which are entirely separated from each other and imperfectly interlocked. In such rock, vertical walls may require lateral support
Crushed but chemically intact rock	Has the character of crusher run. If most or ll of the fragments are as small as fine sand grains and no recementation has taken place, crushed rock below the water table exhibits the properties of a water-bearing sand
Squeezing rock	Slowly advances into the tunnel without perceptible volume increase. A prerequisite for squeeze is a high percentage of microscopic and sub-microscopic particles of micaceous minerals or clay minerals with a low swelling capacity
Swelling rock	Advances into the tunnel chiefly on account of expansion. The capacity to swell seems to be limited to those rocks that contain clay minerals such as montmorillonite, with a high swelling capacity

RQD

In 1966 Deere and Miller developed the Rock Quality Designation index (RQD) to provide a systematic method of describing rock mass quality from the results of drill core logs. Deere described the RQD as the length (as a percentage of total core length) of intact and sound core pieces that are 4 inches (10 cm) or more in length. Several proposed methods of using the RQD for design of rock tunnels have been developed. However, the major use of the RQD in modern tunnel design is as a major factor in the Q or RMR rock mass classification systems described in the following sub-sections. Readers are referred to Subsurface Investigation Manual (FHWA, 2002).

Q System

On the basis of an evaluation of a large number of case histories of underground excavations, Barton et al. (1974) of the Norwegian Geotechnical Institute proposed a Tunnelling Quality Index (Q) for the determination of rock mass characteristics and tunnel support requirements. According to its developer: "The traditional application of the six-parameter Q-value in rock engineering is for selecting suitable combinations of shotcrete and rock bolts for rock mass reinforcement, and mainly for civil engineering projects". The numerical value of the index Q varies on a logarithmic scale from 0.001 to a maximum of 1,000 and is estimated from the following expression (Barton, 2002):

$$Q = \left[\frac{RQD}{J_n}\right] \times \left[\frac{J_r}{J_a}\right] \times \left[\frac{J_w}{SRF}\right]$$

Where *RQD* is Rock Quality Designation, J_n is joint set number, J_r is joint roughness number, J_a is joint alteration number, J_w is joint water reduction factor, and *SRF* is stress reduction factor. It should be noted that RQD/J_n is a measure of block size, J_r/J_a is a measure of joint frictional strength, and J_w/SRF is a measure of joint stress.

Table 6-2 (6-2.1 through 6-2) gives the classification of individual parameters used to obtain the Tunnelling Quality Index Q for a rock mass. It is to be noted that Barton has incorporated evaluation of more than 1,000 tunnels in developing the Q system.

Rock Mass Rating (RMR) System

Z.T. Bieniawski (1989) has developed the Rock Mass Rating (RMR) system somewhat along the same lines as the Q system. The RMR uses six parameters, as follows:

- Uniaxial compressive strength of rock
- RQD
- Spacing of discontinuities
- Condition of discontinuities

- Groundwater condition
- Orientation of discontinuities

The ratings for each of these parameters are obtained from Table 6-3. The sum of the six parameters becomes the basic RMR value as demonstrated in the following example. Table 6-9 presents how the RMR can be applied to determining support requirements for a tunnel with a 33 ft (10 m) width span.

Estimation of Rock Mass Deformation Modulus Using Rock Mass Classification

The in situ deformation modulus of a rock mass is an essential parameter for design, analysis and interpretation of monitored data in any rock tunnel project. Evaluation of the stress and deformation behaviour of a jointed rock mass requires that the modulus and strength of intact rock be reduced to account for the presence of discontinuities such as joints, bedding, and foliation planes within the rock mass. Since the in situ deformation modulus of a rock mass is extremely difficult and expensive to measure, engineers tend to estimate it by indirect methods. Several attempts have been made to develop relationships for estimating rock mass deformation modulus using rock mass classifications.

The modulus reduction method using RQD requires the measurement of the intact rock modulus from laboratory tests on intact rock samples and subsequent reduction of the laboratory value incorporating the in-situ rockmass value. The reduction in modulus values is accomplished through a widely used correlation of RQD (Rock Quality Designation) with a modulus reduction ratio, EM/EL, where EL represents the laboratory modulus determined from small intact rock samples and EM represents the rock mass modulus. This approach is infrequently used directly in modern tunnel final design projects. However, it is still considered to be a good tool for scoping calculations and to validate the results obtained from direct measurement or other methods.

Based on the back analyses of a number of case histories, several methods have been propounded to evaluate the in situ rock mass deformation modulus based on rock mass classification. The methods are summarised.

Rock Tunnelling Methods

Drill and Blast: When mankind first started excavating underground, the choices of tools were extremely limited – bones, antlers, wood and rocks, along with a lot of muscle power. Exactly when and where black powder was first used has been lost in history but it is generally agreed that progress was quite slow until the early 1800's. Then, in the mid 1800's, Alfred Nobel invented dynamite and we began to make significant progress in excavations for mining, civil, and military applications. For the reader who wants to pursue this interesting topic.

Modern drill and blast excavation for civil projects is still very much related to mining and is a mixture of art and science. The basic approach is to drill a pattern of small holes, load them with explosives, and then detonate those explosives thereby creating an opening in the rock.

The blasted and broken rock (muck) is then removed and the rock surface is supported so that the whole process can be repeated as many times as necessary to advance the desired opening in the rock.

By its very nature this process leaves a rock surface fractured and disturbed. The disturbance typically extends one to two metres into the rock and can be the initiator of a wedge failure as discussed previously. As a minimum this usually results in an opening larger than needed for its service requirement and in the need to install more supports than would be needed if the opening could be made with fewer disturbances.

Controlled Blasting Principles

Explosives work by a rapid chemical reaction that results in a hot gas with much larger volume than that occupied by the explosive. This is possible because the explosive contains both the fuel and the oxidizer. When the explosive detonates, the rapidly expanding gas performs two functions: applying a sharp impulse to the borehole wall (which fractures the rock) and permeating the new fractures and existing discontinuities (which pries the fragments apart). To deliver this one-two punch effectively, the explosive is distributed through the rock mass, by drilling an array of boreholes that are then loaded with explosives and fired in an orderly sequence.

Relief

In order to effectively fragment the rock, there must be space for the newly created fragments to move into. If there is not, the rock is fractured but not fragmented, and this unstable mass will remain in place. Therefore the geometry of the array of boreholes must be designed to allow the fragments to move.

Delay Sequencing

To optimize the relief, internal free faces must be created during the blast sequence. To do this, millisecond delay detonators separate the firing times of the charges by very short lengths of time. Historically, because of scatter in the firing times of pyrotechnic detonators, "long" period delays between holes (on the order of hundreds of milliseconds) have been used in tunnel and underground mining, resulting in blasts that last several seconds. This is changing as more accurate electronic detonators are developed.

Tunnel Blast Specifics

As mentioned, tunnel blasting (like underground mine blasting) differs from surface blasting in that there is usually only one free face that provides relief. To blast some large tunnels, an upper heading is blasted first, and the rest of the rock is taken with a bench blast. Often, though, the whole face is drilled and blasted in one event. An array of blastholes is drilled out using drilling equipment that can drill several holes at once. The pattern of drill holes is determined before the blast, taking into account the rock type, the existing discontinuities in the rock (joints, fractures, bedding planes), and of course the desired final shape of the tunnel.

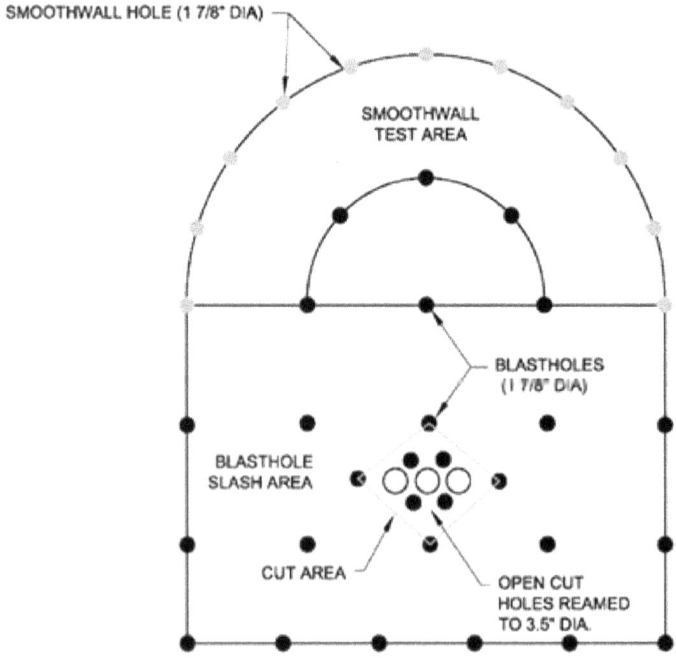

Figure: Example of a Full-face Tunnel Blast

Burn Cut

Because the start of each cut with a solid face has no relief, several extra holes are usually drilled and not loaded with explosives in the immediate neighbourhood of the initiation point. These burn holes are generally larger than the explosively loaded holes, requiring an additional operation beyond the normal drilling. Many different geometries of burn holes are used to optimize the cut, depending on the rock type and joint patterns in a specific tunnel geology. These holes are fired first, with enough firing time to allow the creation of a free face for the following holes to expand into. Production Holes The mass of the rock, following the initiation of the burn cut, are fired in a sequence so that the rock moves in a choreographed sequence, moving into the area opened up by the burn cut, and out into the open space in front of the blast.

Wiring up the charges in the right sequence can be a challenging task in the confined environment of a tunnel. The desired sequence will fire holes so that there is enough time for rock to move out of the way (create relief) but not so much time that the rock surrounding unfired blast holes will fracture (creating a cutoff).

Perimetre Control It is important to blast so that the final wall is stable and as close to the designed location as possible. The final holes are loaded more lightly, and called "perimetre holes" or "smoothwall holes", and fired with some extra delay so that there is sufficient time for rock to fracture cleanly and create little damage to the rock outside of the "neat" line (such damage is called overbreak). Typical blast charges for these smoothwall holes are shown.

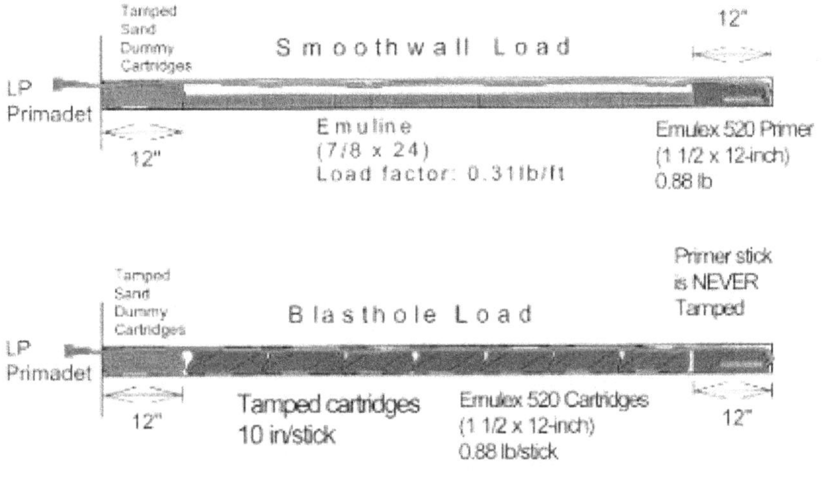

Figure: Typical Blast Charges

Though fired after the Production Holes have been detonated, the Smoothwall Holes are often fired on the same delay period, creating a "zipper" effect of the holes generating a smooth fracture on the perimetre.

Environmental Effects – Vibration and Airblast Not all of the energy from blasting goes into fragmenting rocks – some of it is unavoidably wasted as vibration that propagates away from the blast area. This vibration can be cause for concern both for the stability of the tunnel itself, as well as neighbouring underground and surface structures. Airblast is an air pressure wave that propagates away from the blast site, due to movement of the rock face and also possible venting of explosive gases from the boreholes. This is not so much a problem in tunnelling, where personnel are evacuated from the blast area before a blast, but still must be taken into account.

Blasting – Art vs. Science

Explosives have been used for a long time to excavate rock. With the passage of time, engineers have studied the scientific relationships between the properties of explosives, the controllable variables such as the geometry of a blast and the timing, and uncontrollable variables such as variations in rock type and existing jointing and fracturing. Many relationships can show the most appropriate configuration of the blastholes, timing, and explosive type.

Figure: Drilling for a Tunnel Blast

However, as can be seen from the Figure of actual drilling for a tunnel blast, the ideal is difficult to achieve. Holes are marked out with spray paint on an irregular surface, and drilled in a dirty, often wet environment. The roof is supported with rock bolts and meshes. Lighting is limited. Overall, this makes for a very challenging work environment. Experience, or the "art" of blasting, comes into play in implementing the desired blast design. Choice of an experienced and capable blasting contractor, as well as a blast consultant to advise the contractor, is important to obtain the desired results.

Tunnel Boring Machines (TBM)

While progress and mechanization continued to be applied to drill and blast excavation well into the 1960's, the actual advance rates were still quite low, usually measured in feet per day. Mechanized tunnelling machines or tunnel boring machines had been envisioned for over a century but they had never proven successful. That began to change in the 1960's when attempts were made to apply oil field drilling technology. Some progress was made, but it was slow because the physics were wrong – the machines attempted to remove the rock by grinding it rather than by excavating it. All of that changed in the later 1960's with the introduction of the disk cutter. The disk cutter causes the rock to fail in shear, forming slabs (chips) of rock that are measured in tens of cubic inches rather than small fractions of a cubic inch. Much of the credit for this development, which now allows tunnels to advance at 10's or even 100's of feet per day, belongs to The Robbins Co. Today, tunnel boring machines (TBM) excavate rock mass in a form of rotating and crushing by applying enormous pressure on the face with large thrust forces while rotating and chipping with a number of disc cutters mounted on the machine face (cutterhead). Design of disc cutters RPM, geometry, spacing, thrust level, etc. are beyond the scope of this manual.

Machine Types and Systems

Tunnel Boring Machines (TBMs) nowadays are full-face, rotational (with cutter heads) excavation machines that can be generally classified into two general categories: Gripper and Segment as shown. There are three general types of TBMs suitable for rock tunnelling including Open Gripper/Main Beam, Closed Gripper/Shield, and Closed Segment Shield.

The open gripper/beam type of TBMs are best suited for stable to friable rock with occasional fractured zones and controllable groundwater inflows. As shown in Appendix D, three common types of TBMs belong to this category including Main Beam, Kelly Drive, and Open Gripper (without a beam or Kelly).

The closed shield type of TBMs for most rock tunnelling applications are suitable for friable to unstable rocks which cannot provide consistent support to the gripper pressure. The closed shield type of TBMs can either be advanced by pushing against segment, or gripper. Note that although these machines are classified as a closed type of machine, they are not pressurized at the face of the machine thus cannot handle high external groundwater pressure or water inflows. Shielded TBMs for rock tunnelling include: Single Shield, Double Shield, and Gripper Shield.

Pressurized-face Closed Shield TBMs are predominantly utilised in tunnelling in soft ground. Appendix D presents descriptions for various types of TBMs.

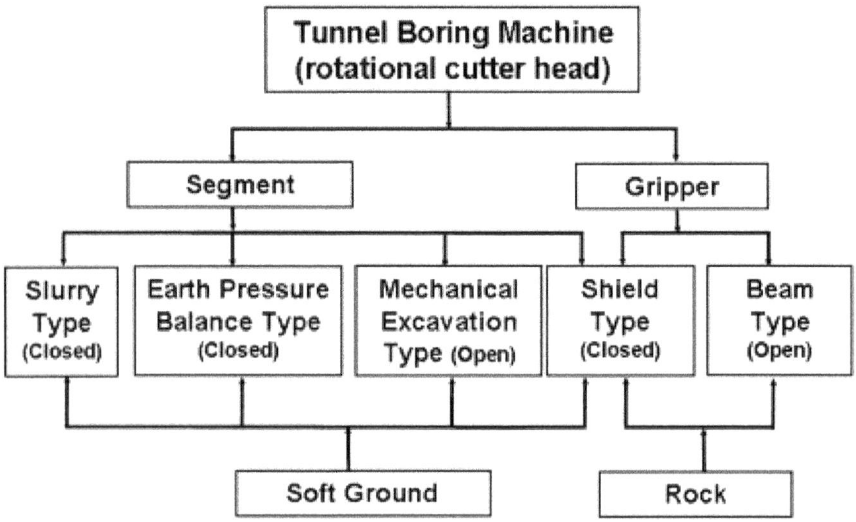

Figure: Classification of Tunnel Excavation Machines

Machine Main and Support Elements

A TBM is a complex system with a main body and other supporting elements to be made up of mechanisms for cutting, shoving, steering, gripping, shielding, exploratory drilling, ground control and support, lining erection, spoil (muck) removal, ventilation and power supply. As shown, the main body of a typical rock TBM (either open or closed) includes some or all of the following components:

- Cutterhead and Support
- Gripper (Except Single Shield TBM)
- Shield (Except Open TBM)
- Thrust Cylinder
- Conveyor
- Rock Reinforcement Equipment

In addition, the main body of a TBM is supported with a trailing system for muck and material transportation, ventilation, power supply, etc. A fully equipped TBM can occupy over 1000 ft of tunnel.

Compatible Ground Support Elements

The use of hard rock TBMs, especially if the TBMs are manufactured specifically for the project.

- Rock reinforcement by roof bolting
- Spiling/forepoling
- Pre-injection
- Steel ring beams with or without lagging (wire mesh, timber, etc.)
- Invert segment
- Shotcrete
- Precast concrete segmental lining
- Others

TBM Penetration Rate

With a rock TBM, the penetration rate is affected by the following factors (from Robbins, 1990):

- Total machine thrust
- Cutter spacing
- Cutter diametre and edge geometry
- Cutterhead turning speed (revolutions per minute)
- Cutterhead drive torque
- Diametre of tunnel
- Strength, hardness, and abrasivity of the rock
- Jointing, weathering and other characteristics of the rock.

However, penetration rate (an instantaneous parameter) by itself does not assure a high average advance rate. The latter requires a good combination of penetration rate and actual cutting time. In turn, actual cutting time is affected by the following factors:

- Learning (start-up) curves
- Downtime for changing cutters
- Downtime for other machine repairs/maintenance
- Overly complex designs

- Back-up (trailing) systems
- Tunnel support requirements
- Muck handling
- Water handling
- Probe hole drilling, grouting
- Available time (total and shift)

The bottom line is that actual utilisation typically runs in the range of 50% as shown.

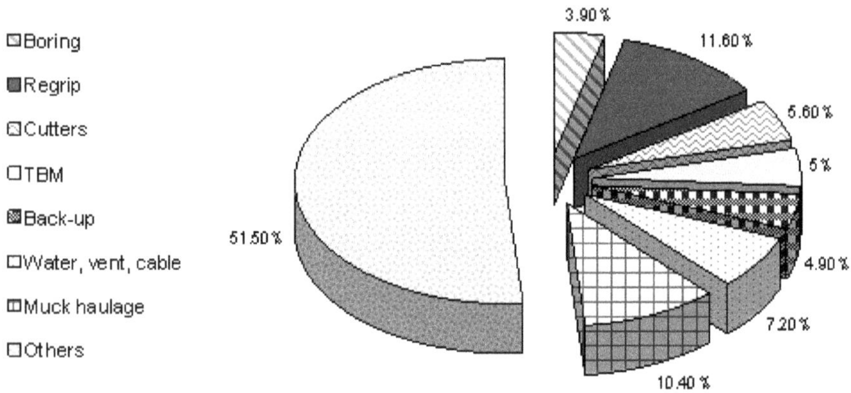

Figure: TBM Utilisation on Two Norwegian Tunnels (After Robbins, 1990)

Rock Tunnelling

Roadheaders: Typical TBM's cut circular tunnels which are the most practical cross section but not always the cross section that provides the most useable volume as a proportion of the total volume excavated. The Japanese and others have developed specialised machines with multiple heads that cut "slots" or other shapes that can be more efficient in providing useable volume.

Another approach to cutting an opening closer to some actual required section is the roadheader. The basic cutting tool for a roadheader is a very large milling head mounted on a boom, which boom, in turn, is mounted on tracks or within a shield. A large size roadheader. Corners must be cut to the curvature of the milling head, but the rest of the walls, crown and invert can be cut to almost any desired shape. In addition and in contrast to a TBM, a single roadheader can cut variable or odd shapes that otherwise would require TBM excavation in combination with drill and blast or drill and blast itself. Because of their adaptability, availability (a few months rather than a year or longer), and lower cost, roadheaders also are the method of choice for relatively short tunnels, say less than one mile in length.

On the negative side, roadheaders are far less efficient on longer drives and in hard rock. The picks on the roadheader are something like 10% as efficient as TBM disks at removing rock, must be replaced very frequently and simply may not be effective in rock with an unconfined compressive strength greater than 20,000 psi (140MPa). Changes and improvements in roadheader design are on-going, however, and it is expected that this will result in constant improvements in these limitations.

The following is a general list when roadheaders may be considered:

- Rock strength below about 20000 - preferably below 15000.
- Short runs, one of a kind openings
- Odd, non-circular shapes
- Connections, cross passages, etc
- Low to moderate abrasivity
- Preferably self supporting rock
- No or small inclusions - chert etc
- Nominal water pressure

Other Mechanized Excavation Methods

Other mechanized excavation methods are being developed by specialised equipment manufacturers to address specific issues in mining and civil applications. A good example of such developments is the "Mobile Miner" developed by Robbins as a "non-circular hard rock cutting system to be applied to underground mine development" (Robbins, 1990). The mobile miner is described as follows:

"A boom mounts a large cutter wheel with a transverse axis having rows of cutters arranged only on the periphery of the wheel. As the boom is swung from side to side an excavated shape is generated with a flat roof and floor and curved walls. Although the prototype machines have operated only with this side-swinging action, in order to cut openings which are better suited to vehicular tunnels the cutterhead boom must be elevated up and down and at the same time swung from side to side. In this way a horseshoe-shaped excavation can be generated.

To date such machines have had some success in excavating openings approaching a horseshoe or slot configuration, but they are not commonly used. However, they do illustrate the points that shapes other than circular can be cut and that inventive and special-purpose machines are constantly being developed.

Sequential Excavation Method (SEM)/ New Austrian Tunnelling Method (NATM)

In actual practice, the Sequential Excavation Method (SEM)/New Austrian Tunnelling Method (NATM) has been adapted from its original concept, which applied to rock tunnels only, to a more general concept that applies to tunnels in either soil or rock.

Types of Rock Reinforcement and Excavation Support

Excavation Support Options

The purpose of an initial support (sometimes called temporary lining, or temporary support of excavation) in rock tunnelling is to keep the opening open, stable and safe until the final lining is installed and construction is complete. As a consequence the initial support system in a rock tunnel can be one or a combination of a number of options:

- Rock reinforcement (i.e., rock dowels, rock bolts, rock anchors, etc.)
- Steel ribs
- Wood or other lagging
- Lattice girders
- Shotcrete
- Spiles or forepoling
- Concrete
- Re-steel mats
- Steel mats
- Cables
- Precast concrete segments
- Others

The first five above are the most common on US projects, and of those, a combination of rock bolts or dowels and shotcrete is the single most common. Especially in good (or better) rock tunnels, modern rock bolting machines provide rapid and adjustable "support" close to the heading by knitting and holding the rock (ground) arch in place, thus taking maximum advantage of the rock's ability to support itself. Preferably, shotcrete is added (if needed) a diametre or so behind the face where its dust and grit and flying aggregate is not the problem for both workers and equipment that it is at the heading. Where there is a concern with smaller pieces of rock falling, the system can be easily modified by adding shotcrete closer to the face or more usually, by embedding any of a number of types of steel mats in the shotcrete.

Where the rock quality is lower there is currently a movement toward replacing steel ribs by lattice girders – perhaps somewhat more so in Europe than in the US. Like steel ribs, the lattice girders form a template of sorts for the shotcrete and for spiling. However, the lattice girders are lighter and can be erected faster. To provide the same support capacity, the lattice girder system may require nominally more shotcrete (e.g., an additional ½ to 1 inch) but that is more than compensated for by the easier and faster erection. A second new trend is the use of steel fibre reinforced shotcrete. The fibre doesn't change the compressive strength significantly but does produce a significant increase in the toughness or ductility of the shotcrete.

.Rock Reinforcement

Rock reinforcement including rock dowels and bolts are used to hold loose (key) blocks in place and/or to knit together the rock (ground) arch that actually provides the support for an opening in rock. Dowels and bolts are very similar but the differences in their behaviour can be quite significant.

Rock Dowel

Rock dowels as shown, are passive reinforcement elements that require some ground displacement to be activated. Similar to passive concrete reinforcement, the reinforcement effect of dowels is activated by the movement of the surrounding material. In particular, when displacements along discontinuities occur, dowels are subject to both shear and tensile stresses. The level of shear and tensile stress and the ratio between them occurring during a displacement is dependent on the properties of the surrounding ground, the properties of the grout material filling the annular gap between the dowel and the ground and the strength and ductility parameters of the dowel itself. Also, the degree of dilation during shear displacement influences the level of stress acting within the dowel. Table describes various types of rock dowels. In addition, Table summarises commonly used rock dowels and application considerations for the installation as part of initial support in SEM tunnelling in rock.

For example a #9 dowel 10 ft. long will have to elongate almost 0.2 inch before it develops its full design capacity of 40,000 lb. This may not be a concern in most applications where there is some interlocking between rock blocks due to the natural asperities on discontinuity surface.

Ribs and Lagging

Ribs and lagging are not used as much now as they were even a couple of decades ago. However, there are still applications where their use is appropriate, such as unusual shapes, intersections, short starter tunnels for TBM, and reaches of tunnel where squeezing or swelling ground may occur.

In 1946, Proctor and White (with major input from Dr. Karl Terzaghi) wrote the definitive volume "Rock Tunnelling with Steel Supports". Their design approach assumes the ribs are acted upon by axial thrust and by bending moments, the latter a function of the spacing of the lagging or blocking behind the ribs. This approach is still valid when wood or other blocking is used with steel ribs and hence will not be repeated here. In today's applications, steel ribs are often installed with shotcrete being used instead of wood for the blocking (lagging) material. When shotcrete is used, it often does not fill absolutely the entire void between steel and rock. Hence, with properly applied shotcrete it is recommended that the maximum blocking point spacing be taken as 20 in. and the design proceed according to the Proctor and White procedure.

Shotcrete

Shotcrete is simply concrete sprayed into place through a nozzle. It contains additives to gain strength quicker and to keep it workable until it is sprayed. Shotcrete can be made with or without the addition of reinforcing fibers and can be sprayed around and through reinforcing bars or lattice girders. Both the quality and properties of shotcrete can be equal to those of cast in place concrete but only if proper care and control of the total placement procedure is maintained throughout.

Lattice Girder

Lattice girders are support members made up of steel reinforcement bars laced together (usually) in a triangular pattern and rolled to match the shape of the opening. Because their area is typically very small compared to the surrounding shotcrete, lattice girders do not, by themselves, add greatly to the total support of an opening. However, they do provide two significant benefits:

1. They are typically spaced similarly to rock bolts, thus they quickly provide temporary support to blocks having an immediate tendency to loosen and fall
2. They provide a ready template for assuring that a sufficient thickness of shotcrete is being applied

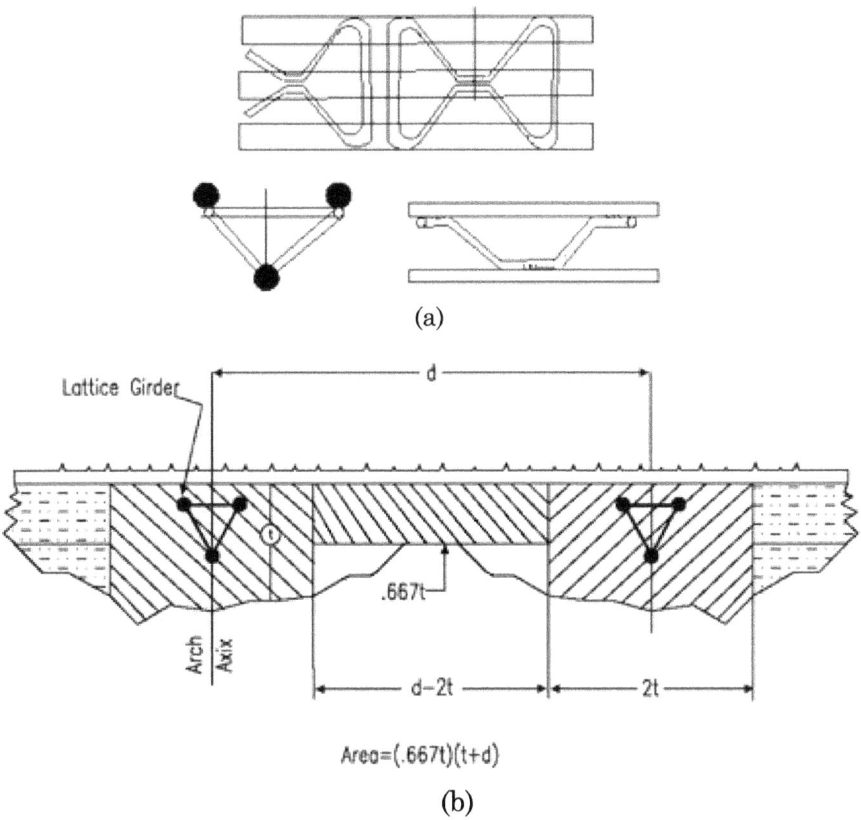

(b)

Figure: (a) Lattice Girder Configuration (Emilio-2-2901-1997); (b) Estimation of Cross Section for Shotcrete-encased Lattice Girders (Emilio-2-2901-1997)

Generally, lattice girders are used much more frequently in tunnels driven by the sequential excavation method.

Spiles and Forepoles

Spiles and forepoles are used interchangeably to describe support elements consisting of pipes or pointed boards or rods driven ahead of the steel sets or lattice girders. These elements (herein called spiles) provide temporary overhead protection while excavation for and installation of the next set or girder is accomplished. Typically, spiles are driven in an overlapping arrangement as shown, so that there is never a gap in coverage. Design of spiles is best described as "intuitive" as it must be kept flexible and constantly adjusted in the field as the ground behaviour is observed during the construction. A working first approximation of design load might be a height of rock equal to 0.1B to 0.25B, where B is the width of the opening. For pre-support measures involving spiling or grouted pipe arch canopies that bridge over the unsupported excavation round.

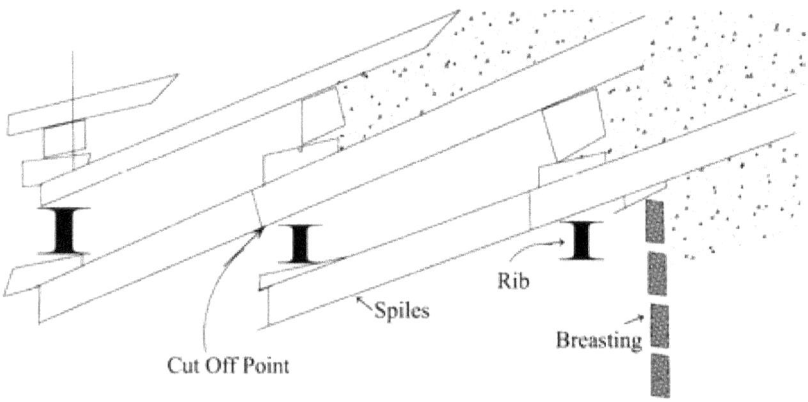

Figure: Spiling (Forepoling) Method of Supporting Running Ground

Precast Segment Lining

Tunnel lining consisting of precast segments may be used in single-pass or double pass lining systems to support rock loadings and water pressures. Generally the concrete segments are reinforced either with reinforcement bar or fibre. The segment ring usually consists of five to seven segments with a key segment. The ring division and the segment dimension have to be optimized according to project specific requirements such as tunnel diametre, maximum size for transport and installation (erector), and number of thrust jacks and their distribution over the range of the ring.

The precast segment concrete lining is mostly used in TBM tunnel construction projects and, at this time, more frequently in soft ground tunnels. The segmental ring is erected in the TBM tail shield and during the advance, the rams act on the ring. The ring never can be independent from the TBM, hence the design of the TBM and the segmental ring must be harmonized. Rams must act on prepared sections of the ring, rolling of the tunnel shield and the ring must be taken into account. The ram axis should be identical with the ring axis. The ring taper should be designed according to the TBM curve drive capabilities and not only according to the designed tunnel axis.

Design and Evaluation of Tunnel Supports

There exists a wide range of tunnel support systems as shown in previous sections. In recent years the tunnelling community has moved away from support to reinforcement as the basic approach. That is, from providing heavy structures, primarily ribs and lagging, to using rock bolts and dowels, spiling, lattice girders and shotcrete. In all of these latter systems the goal is to keep the rock from moving and blocks from loosening thereby keeping a large dead load of rock from coming onto the support system; that is holding the rock together and causing the ground around the opening to form a natural and self supporting ground arch around the opening.

This trend was started in the U.S. on the Washington D.C. subway system where on two successive sections the amount of structural steel support was reduced by three quarters. This was accomplished by holding the rock in place with rock bolts until the final lining of shotcrete with light ribs at four foot centres could be installed and become effective in causing the rock to help support itself. In contrast, the previous section relied upon steel ribs to carry the dead (rock) loads and thus required twice the weight of steel members at one-half the spacing. Tunnel support design is an iterative process including assumptions on support type installed and evaluating the support pressure it provides. Typical tunnel support systems used in the current practice for various ground conditions. This table can be used for the initial selection of the support system to initiate the interaction and iteration.

Table: Typical Initial Support and Lining Systems Used in the Current Practice

Ground	*Rock bolts*	*Rock bolts with wire mesh*	*Rock bolts with shotcrete*	*Steel ribs and lattice girder with shotcrete*	*Cast-in-place concrete*	*Concrete segments*
Strong	O	O				
rock		O	O			
		O	O	O		
Medium			O	O	O	
Rock				O	O	O
				O	O	O
Soft Rock				O	O	O
Soil						

In making the selection of support measures for a given project, however, the full range of possible support system should be considered simply because each project is unique. Factors to be considered include the following:

1. *Local custom:* contractors like to use systems with which they are familiar.
2. *Relative costs:* for example, is it cost effective to design bolts with suitable corrosion resistance to assure their permanence.
3. Availability of materials

This Section introduces design practice and evaluation of initial tunnel supports, including empirical, analytical and numerical methods. Design of underground structures can be based largely on previous experience and construction observations to assess expected performances of specified ground support systems.

Empirical Method

Terzaghis's tunnelman's classification of rock condition and recommended rock loadings, expressed as a function of tunnel size are presented in Table 6-8. These recommendations sprang from Terzaghi's observations in the field and his trap door experiments in the laboratory.

Table 2: Suggested Rock Loadings from Terzaghi's Rock Mass Classification

Rock condition	***Rock Load, H_p (ft)***	***Remarks***
Hard and intact	Zero	Light lining, required only if spalling or popping occurs
Hard stratified or schistose	0 to 0.5 B	Light support. Load may change erratically from point to point
Massive, moderately jointed	0 to 0.25 B	
Moderately blocky and seamy	0.25B to 0.35 (B + H_t)	No side pressure
Very blocky and seamy	(0.35 to 1.10) (B + H_t)	Little or no side pressure
Completely crushed but chemically intact	1.10 (B + H_t)	Considerable side pressure. Softening effect of seepage towards bottom of tunnel requires either continuous support for lower ends of ribs or circular ribs
Squeezing rock, moderate depth	(1.10 to 2.10) (B + H_t)	Heavy side pressure, invert struts required. Circular ribs are recommended
Squeezing rock, great depth	(2.10 to 4.50) (B + H_t)	
Swelling rock	Up to 250ft. irrespective of value of (B + H_t)	Circular ribs required. In extreme cases use yielding support

As a first approximation to rock bolt or dowel selection, Cording et al. (1971) provides a compilation of case histories for underground rock excavations based on excavation sizes (span width and height):

- A crown support pressure equal to a rock load having a height of 0.3B.

- A sidewall support pressure of 0.15H.
- A crown bolt length of 0.33B.
- A sidewall bolt length of 0.33H.

where B is the opening width. In rock with an RQD greater than 75%, it is expected that the sidewall pressure typically will be smaller (often zero) than estimated above and only spot bolts to hold obvious wedges will be required. It should be noted that "the Q-system has its best applications in jointed rock mass where instability is caused by rock falls.

Table 3: Guidelines for Excavation and Support of 10 m Span Rock Tunnels in Accordance with the RMR System

Rock mass class	***Excavation***	***Rock bolts (20 mm diameter, fully grouted)***	***Shotcrete***	***Steel sets***
I – very good rock RMR: 81-100	Full face 3 m advance	Generally no support required except spot bolting		
II – Good rock RMR: 61-80	Full face, 1-1.5 m advance. Complete support 20 m from face	Locally, bolts in crown 3 m long, spaced 2.5 m with occasional wire mesh	50 mm in crown where required	None
III – Fair rock RMR: 41-60	Top heading and bench 1.5-3 m advance in top heading. Commence support after each blast. Complete support 10 m from face	Systematic bolts 4 m long spaced 1.5-2 m in crown and walls with wire mesh in crown	50-100 mm in crown and 30 mm in sides	None
IV – Poor rock RMR: 21-40	Top heading and bench 1.0-1.5 m advance in top heading. Install support concurrently with excavation, 10 m from face	Systematic bolts 4-5 m long, spaced 1-1.5 m in crown and walls with wire mesh	100-150 mm in crown and 100 mm in sides	Light to medium ribs spaced 1.5 m where required
V – Very poor rock RMR: < 20	Multiple drifts 0.5-1.5 m advance in top heading. Install support concurrently with excavation. Shotcrete as soon as possible after blasting.	Systematic bolts 5.6 m long, spaced 1-1.5 m in crown and walls with wire mesh. Bolt invert.	150-200 mm in crown, 150 mm in sides, and 50 mm on face	Medium to heavy ribs spaced steel lagging and forepoling if required. Close invert.

Note: Above table assumes excavation by drill and blast.

For most other types of ground behaviour in tunnels the Q-system, like most other empirical (classification) methods has limitations. The Q support chart gives an indication of the support to be applied, and it should be tempered by sound and practical engineering judgment". Also note that the Q-system was developed from over 1000 tunnel projects, most of which are in Scandinavia and all of which were excavated by drill and blast methods. When excavation is by TBM there is considerably less disturbance to the rock than there is with drill and blast. Based upon study of a much smaller data base, Barton (1991) recommended that the Q for TBM excavation be increased by a factor of 2 for Qs between 4 and 30.

Barton et al. (1980) proposed rock bolt length, maximum unsupported spans and roof support pressures to supplement the support recommendations. The length of rock bolts, L can be estimated from the excavation width, B and the Excavation Support Ratio (ESR) as follow:

$$L = \frac{0.15B}{ESR}$$

The maximum unsupported span can be estimated from:

Maximum Unsupported Span = $2\ ESR\ Q^{0.4}$, in metres

Grimstad and Barton (1993) proposed a relationship between Q value and the permanent roof support pressure, P_{roof} as follow:

$$P_{roof}(MPa) = \frac{2\sqrt{J_n}Q^{1/3}}{3J_r}$$

The value of Excavation Support Ratio (ESR) is related to the degree of security which is demanded of the support system installed to maintain stability of the excavation. Barton et al. (1974) suggested ESR values for various types of underground structures. An ESR value of 1.0 is recommended for civil tunnel projects.

Table 4: Excavation Support Ratio (ESR) Values for Various Underground Structures

	Excavation Category	Suggested ESR Value
A	Temporary mine openings	3 – 5
B	Permanent mine openings, water tunnels for hydro power (excluding high pressure penstocks), pilot tunnels, drifts and headings for large excavations.	1.6
C	Storage rooms, water treatment plants, minor road and railway tunnels, surge chambers, access tunnels	1.3
D	Power stations, major road and railway tunnels, civil defense chambers, portal intersections	1.0
E	Underground nuclear power stations, railway stations, sports and public facilities, factories	0.8

Pre-Support and Other Ground Improvement Methods

Pre-support is used in both rock and soil tunnels, perhaps somewhat more frequently in soil tunnels. In rock tunnel applications pre-support may be called for when the tunnel encounters zones of badly weathered and/or broken rock. In such rock, the stand up time may be too short to install the usual support system.

Pre-support may include a number of techniques. For example, spiles and forepoling typically are installed through and ahead of the tunnel face. These members are driven or drilled as shown schematically and pass over the support nearest the face and under or through the next support back from the face. Thus, overlapping "cones" of spiles are formed and this results in a sawtooth pattern to the opening profile. These spiles are usually selected based upon experience and judgment as there is no known design method. Therefore, successful application usually rests on the workers in the field because they are at the face and have to make the decisions in real time and in short time as the ground is exposed and its behaviour observed.

Sequencing of Excavation and Initial Support Installation

The three principal excavation methods for rock tunnels are as follows:

- Drill and blast (including SEM/NATM) for full face or multiple heading advance of any shape in any rock.
- Roadheader for full face or multiple heading advance of a shape in rock up to moderate strength.
- TBM for full face (generally round only) in any rock.

When an excavation is made in intact rock by any method there is an adjustment (or redistribution) in the stresses and strains around that excavation. This adjustment, however, quickly dissipates such that the change is only about six percent at a clear distance of three radii from the wall of the opening. The insitu stresses in the rock are generally low for most highway tunnels because those tunnels are at relatively shallow depth. Thus, in intact rock the ("elastic") stresses resulting from this redistribution do not exceed the rock strength so stability is not a concern.

However, rock in reality is a jointed (blocky) material and it is the behaviour of a blocky mass that nearly always governs the behaviour of the tunnel. Evert Hoek describes this behaviour as follows (Hoek, 2000): "In tunnels excavated in jointed rock mass at relatively shallow depth, the most common types of failure are those involving wedges falling from the roof or sliding out of the sidewalls of the openings.

These wedges are formed by intersecting structural features, such as bedding planes and joints, which separate the rock mass into discrete but interlocked pieces. When a free face is created by the excavation of the opening, the restraint from the surrounding rock is removed. One or more of these wedges can fall or slide from the surface if the bounding planes are continuous or rock bridges along the discontinuities are broken.

Unless steps are taken to support these loose wedges, the stability of the back and walls of the opening may deteriorate rapidly. Each wedge, which is allowed to fall or slide, will cause a reduction in the restraint and the interlocking of the rock mass and this, in turn, will allow other wedges to fall. This failure process will continue until natural arching in the rock mass prevents further unravelling or until the opening is full of fallen material.

The steps which are required to deal with this problem are:

Step 1: Determination of average dip and dip direction of significant discontinuity sets.

Step 2: Identification of potential wedges which can slide or fall from the back or walls.

Step 3: Calculation of the factor of safety of these wedges, depending upon the mode of failure.

Step 4: Calculation of the amount of reinforcement required to bring the factor of safety of individual wedges up to an acceptable level."

The concepts for and applications of sequencing of excavation and initial support installation are generally based on drill and blast excavation, but also apply to roadheader excavation. These concepts can be summarised in "one sentence" as follows: do not excavate more than can be quickly removed and quickly supported so that ground control is never compromised.

Face Stability

In general, face stability is not as great a concern in rock tunnels as in soil tunnels because the rock stresses tend to arch to the sides and ahead of the face. However, in low strength rock, in areas where the rock is broken up or where the rock is extremely weathered face stability may be an issue. The secret to successful tunnelling where face stability may be an issue is to assure that individual headings are never so large that they cannot be quickly excavated and quickly supported. In addition, where groundwater exists it should be drawn down or otherwise controlled because, as noted by Terzaghi, unstable ground is usually associated with or aggravated by groundwater under pressure.

Surface Support

Surface support in a rock tunnel may be supplied by ribs and lagging as discussed above, or, more frequently now, by shotcrete in combination with rock bolts or dowels, steel sets, lattice girders, wire mesh or various types of reinforcement mats. For the most part modern rock tunnels are supported by shotcrete and either rock bolts or lattice girders.

Either system provides a flexible support that takes advantage of the inherent rock strength but that can be stiffened simply and quickly by adding bolts, lattice girders and/or shotcrete. In addition, lattice girders provide a simple template by which to judge the thickness of shotcrete. For other situations wire mesh or reinforcement mats have proven to successfully arrest and hold local raveling until sufficient shotcrete can be applied to knot the whole system together and hold it until the shotcrete attains its strength.

Ground Displacements

For the most part, ground displacements around a rock tunnel can be estimated from elastic theory or calculated using any of a number of computer programs. Elastic theory allows an approximate calculation of the ground displacements around a round tunnel in rock, as shown. The approximate radial displacement at a point directly around a tunnel in elastic rock is given by:

$$u = \frac{P_{\Sigma}(1+\nu)}{E}\frac{a^2}{r}$$

Where:

u = Radial movement, in.
P_z = Stress in the ground
v = Poisson's ratio
E = Rock mass modules
a = Radius of opening
r = Radius to point of interest as presented.

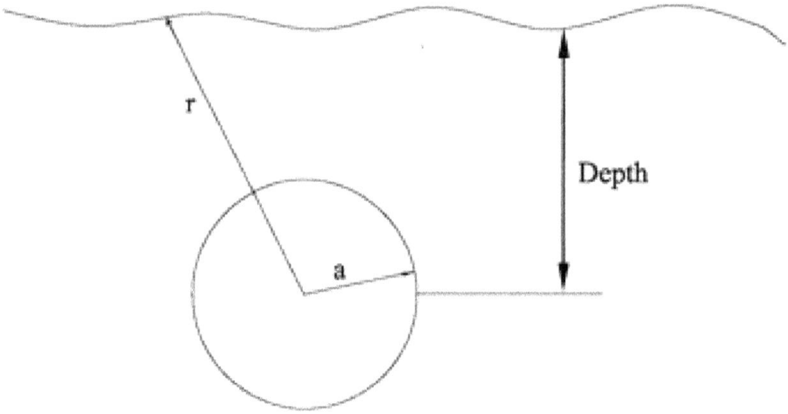

Figure: Elastic Approximation of Ground Displacements around a Circular Tunnel in Rock

For any shape other than circular, one can usually sketch a circle that most nearly approximates the true opening and use the radius of that circle in the above solution for an approximate displacement. However, in the rare case where the precise value of movement might be a concern, then it should be determined by numerical analysis. Displacement contours induced by two tunnel excavation, calculated by Finite Element Method, are presented in figures.

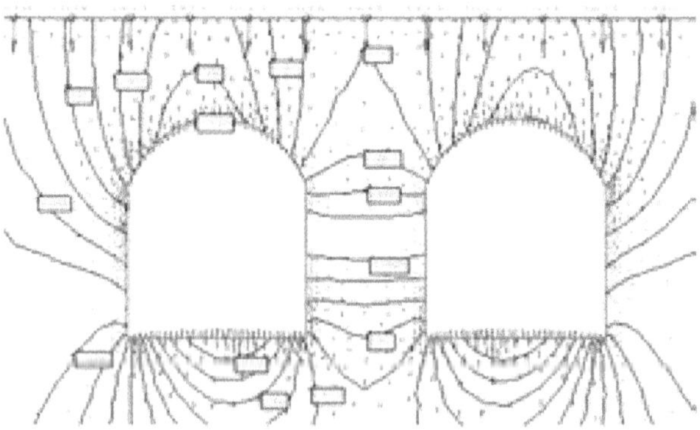

Figure: Ground Displacement Contours Calculated by Finite Element Method

Groundwater Control During Excavation

Groundwater control in rock can take many forms depending on the nature and extent of the problem. In fact, for many cases experience has proven that a combination of control methods may be the best solution. For a given tunnel it may also be found that different solutions apply at different locations along the alignment.

Dewatering at the Tunnel Face

Dewatering at the tunnel face is the most common method of groundwater control. This consists simply of allowing the water to drain into the tunnel through the face, collecting the water, and taking it to the rear by channels or by pumping. It then joins the site water disposal system. Note that if there are hydraulic or other leaks or spills at the TBM or other equipment in the tunnel such contaminants are in this water.

Drainage Ahead of Face from Probe Holes

Probe holes ahead of the tunnel may be placed to verify the characteristics of the rock and hence to provide information for machine operation and control. These holes will also predrain the rock and provide warning of (and drain) any methane, hydrogen sulfide or any other gas, petroleum, or contaminant that may be present. In areas where there are such known deposits of gas or other contaminants it is common (and recommended) practice to keep one or more probe holes out in front of the machine. When such materials are encountered, the probes alert the workers to the need to increase the frequency of gas readings, to increase the volume of ventilation or to take other steps as required to avoid the problem of unexpected or excessive gas in the tunnel.

Drainage from Pilot Bore/Tunnel

Pilot tunnels can provide a number of benefits to a larger tunnel drive, including:

- Groundwater drainage
- Gas or other contaminant drainage
- Exploratory information on the geology
- Grouting or bolting galleries for pre-support of a larger opening
- Rock behaviour/loading information for design of the larger opening

The question of location and size of pilot tunnel always leads to a spirited discussion such that no two are ever the same. They are typically six to eight feet in general size and may be located at one or more of several locations. As one example, on the H-3 project in Hawaii there was concern that huge volumes of water might be encountered. This is a 36 ft + highway through the mountain to the opposite side of the island. Borings were limited, but did not indicate huge volumes of water. However it was common knowledge that similar sites contained water-filled cavities large enough for canoe navigation and there was concern that a similarly large volume would be found in the H-3 tunnel. The pilot tunnel, proved that water was not a major concern and at the same time provided a second, unexpected benefit: by being able to see and analyse the rock for the whole tunnel bore, the winning contractor determined that he could perform major parts of the excavation by ripping with state-of-the-art large rippers in lieu of using drill and blast techniques. Because of this evaluation he was able to shave off millions of dollars in his bid and accelerate the construction schedule by several weeks. As an added benefit, the pilot tunnel was enlarged slightly and now is the permanent access (by way of special drifts) for maintenance forces to access the entire length of tunnel with small pickups without using the active traffic lanes.

Grouting

Groundwater inflow into rock tunnels almost exclusively comes in at joints, bedding planes, shears, fault zones and other fractures. Because these can be identified grouting is the most commonly used method of groundwater control. A number of different grout materials are used depending on the size of the opening and the amount of the inflow.

The design approach is first to detect zones of potentially high groundwater inflow by drilling probe holes out in front of the tunnel face. Second, the zones are characterised and, hopefully, the major water carrying joints tentatively defined. Then, third, a series of grout holes are drilled out to intercept those joints 10 feet to a tunnel diametre beyond the tunnel face or wall. Fourth, using tube-a-machetes, cement and/or water reactive grouts are injected to seal off the water to a level such that succeeding holes are drilled as the fifth step and injected with finer, more penetrating grouts such as micro-fine or ultra-fine cements and/or sodium silicate can be injected to complete the sealing off process. Based on evaluation of the grouting success additional holes and grouting may be required to finally reduce the inflow to an acceptable level. Typically it will be found that steps four and five must be repeated, trial and error, until the required reduction in flow is achieved.

Freezing

On rare occasions, it may become necessary to try freezing for groundwater control in a tunnel in rock. This might occur, for example, at a shaft where it was necessary to control the groundwater locally for a breakout of a TBM into the surrounding rock. If upon beginning excavation of the TBM launch chamber it were found that the water inflow was too great the alternative control methods would be to grout as discussed above or perhaps to freeze.

Any examples in the U.S. where freezing has been used in a rock tunnel, probably for a very simple reason that high inflow encountered into a rock tunnel would be concentrated at the joints present in the rock. The concentration would usually result in a relatively high velocity of flow. Such velocity would typically exceed six feet per day, the maximum groundwater velocity for which it is feasible to perform effective freezing. Thus, for the most part freezing would not be used in rock tunnelling except very locally, as discussed above, and even then it might be necessary to use liquid nitrogen to perform the freezing.

Closed Face Machine

A closed face machine could be used for rock tunnelling in high groundwater flow conditions over short lengths. In reality such a machine would be more like an earth pressure balance (EPB) machine with sufficient rock cutters installed to excavate the rock. For any extended length (greater than a few hundred feet) this would typically be uneconomical. The machine would have to grind up the rock cuttings and mix, the resulting "fines" with large quantities of conditioners and the existing water to result in a plastic material. This is necessary for the EPB to control the face in front of the bulkhead and to bring the material from its pressurized state at the face down to ambient by means of the EPB screw conveyor.

For these reasons, one would not normally plan to build a closed face rock machine but to equip an EPB with rock cutters for driving short stretches in rock within a longer soft ground tunnel. An exception to this general statement would be a rock tunnel in weak or soft rock such as chalk, marl, shale, or sandstone of quite low strength such that it essentially behaved as high strength soft ground.

Other Measures of Groundwater Control

The groundwater control methods discussed above probably account for more than 95% of the cases where such control is required in a tunnel in rock. For the odd tunnel (or shaft) where something else is required the designer may have to rely on experience and or ingenuity to come up with the solution. A few suggestions are given here, but really inventive solutions may have to be developed on a case-by-case basis.

Compressed Air once was a mainstay for control of groundwater or flowing or squeezing ground conditions but it is used very infrequently in modern construction. Where the tunnel (or shaft) can be stabilised by relatively low pressures (say 10 psi or less) it may still be used. However, it requires compressor plants, locks, special medical emergency preparation and decompression times.

Panning may be attractive in some cases where the water inflow is not too excessive and is concentrated at specific points and/or seams. In this case pans are placed over the leaks and shotcreted into place. Water is carried in chases or tubes to the invert and dumped into the tunnel drainage system

Drainage Fabric is now frequently used in rock tunnels. These geotechnical fabrics can be put in over the whole tunnel circumference or, more often, in strips on a set pattern or where the leaks are occurring. Fastened to the surface of the rock with the waterproof membrane portion facing into the tunnel, this fabric is then sandwiched in place by the cast-in-place concrete lining. The fibrous portion of the fabric provides a drainage pathway around and down the tunnel walls and into a collection system at the tunnel invert.

Permanent Lining Design Issues

For many tunnels the principle purpose of the final lining is to prepare the tunnel for its end use, for example, to improve its aesthetics for people or its flow characteristics for water conveyance. Thus, the final lining may consist of cast-in-place concrete, precast concrete panels, or shotcrete.

On the Washington DC subway for example both cast-in-place concrete and precast concrete panels were used. For downtown stations a variety of initial support schemes were used but a final lining of cast-in-place concrete, with a "waffle" interior finish was used for final support and lining. For outlying stations both initial support and the final structural lining were provided by rock bolts, embedded steel sets and shotcrete all installed as the stations were excavated. The precast concrete segmental inner lining (with waffle finish) that was installed at the outlying stations is architectural only – it carries no rock load. Precast concrete segments are more common in soil than in rock tunnels because in soil they are both initial and (sometimes) final support and they provide the reaction for propelling the machine forward. In rock tunnels the machine typically propels itself by reaction against grippers set against the rock.

Rock Load Considerations

Rock loads can be evaluated empirically or analytically. The calculated rock loads are often times described as roof load, side load, and eccentric load, where roof load and side load are uniformly distributed. It is recommended that the permanent lining is designed based on the uniform loads (roof and side loads) and checked by eccentric load case.

The question of what "loads" to use for design of the permanent lining of a tunnel in rock always raises interesting challenges. Fundamentally, three conditions are possible:

1. If initial support(s) are installed early and correctly, it can be showned, that they will not deteriorate within the design life of the structure, and if the opening is stable, then a structural final lining is not required.
2. If initial support(s) are installed early and correctly, the opening is stable (with no continuing loosening), but it cannot be demonstrated that the initial supports will remain completely effective for the design life of the structure, then the load(s) on the final lining may be essentially equal to those of the initial support. An example of this situation is the H-3 tunnel in Hawaii where initial support is provided by 14-ft rock bolts and the load on the final lining was assumed to be 14 feet of rock, analysed in three ways:
    - o Uniform load across the entire tunnel width
    - o Uniform load across half of the tunnel width
    - o Triangular load across the entire tunnel width with the maximum at the centreline.
3. If initial support(s) are providing a seemingly stable opening but it is known that additional support is required for long term stability then that support must be provided by the final lining. An example of this situation is the Superconducting Super Collider where tunnels in chalk were initially stabilised by pattern rock bolts in the crown and spot bolts elsewhere. Months later, however, slaking (and perhaps creep) resulted in linear wedges (with dimensions up to approximately by one-third the tunnel diametre) "working" and sometimes falling into tunnels driven and supported months earlier. To be stable long-term, a lining or additional permanent rock bolts capable of supporting these wedges or blocks would have been necessary.

Figure: Unlined Rock Tunnel in Zion National Park, Utah

As illustrated by the above, determination of the requirement for and value of "loads" to be used for design of final linings in tunnels cannot be prescribed in the manner that is possible for structural beams and columns. Rather, the vagaries of nature must be understood and applied by all on the design and construction team.

Groundwater Load Considerations

For conventional tunnels, the groundwater table is lowered by tunnel excavation, because the tunnels act as a drain. When the undrained system is considered, the groundwater lowering measures are disrupted after the final lining is placed and the groundwater table will reestablish its original position. For a drained system, the groundwater is lowered and will lowered so long as rainfall or at the project site seepage is not sufficient to raise the groundwater table. For underwater tunnels, the groundwater table keeps constant due to the water body above the tunnels and full hydrostatic water pressure should be considered with an undrained system unless an intensive grouting program is implemented in the surrounding ground.

This section discusses factors affecting groundwater flow regime and interaction with concrete lining, and methods to estimate groundwater loadings in the lining design including empirical method, analytical solution, and numerical method.

Factors on the Lining Loads due to Water Flow

Groundwater loadings on the underwater tunnel linings can be reduced with a drained system while the groundwater table keeps constant. The main factors that affect water loads on the underwater tunnel linings due to water flow are: (1) relative ground-lining permeability; (2) relative ground-lining stiffness; and (3) geometric factors such as depth below the water body. The water loads on the lining are greatly dependent on the relative permeability between the lining and surrounding ground. For a tunnel where the lining has a relatively low permeability when compared to the surrounding ground, the lining will behave almost as impermeable and almost no head will be lost in the surrounding ground resulting in hydrostatic water pressures applied directly on the lining.

A relatively permeable lining, on the other hand, will behave as a drain and almost no head will be lost when the water flows through the lining and no direct loads will act on the lining. The loads due to the groundwater will only act on the lining indirectly through the loads applied by the seepage force onto the surrounding ground.

The influence of the relative stiffness is well visualized for the tunnels in a stiff rock mass, where the linings are not designed for the full hydrostatic water pressures by using drained systems. For tunnels in soft ground, the linings are normally designed to withstand a full hydrostatic load.

Empirical Groundwater Loads

The empirical groundwater loading conditions used for the design of tunnel linings in New York are shown and are based on empirical data. As indicated in figure, the groundwater loading diagram follows hydrostatic pressure to a maximum near the tunnel springline (head of Hs), is held constant over a sidewall area of 1/3Hsw, then decreases to 10 percent of hydrostatic pressure at the invert (0.1Hw).

The empirical loads shown, are based on the assumptions that the drainage system is to be comprised of a wall drainage layer (filter fabric), invert drainage collector pipes placed behind the wall and below the cavern floor, and a drainage blanket developed by covering the entire invert with a gravel layer. The water load at the invert level is reduced to 10 percent of hydrostatic water pressure at the invert level with a well-sized and designed gravel drainage bed and drain pipe(s) in the invert (including appropriate provisions and follow up actions for long term maintenance). Under other circumstances, 25 percent of hydrostatic water pressure is recommended at the invert level. The empirical loads are probably conservative but address concerns that groundwater percolating through the wall rock over time could possibly clog the drainage layer (fabric) placed outside the concrete wall causing a buildup of groundwater pressure beyond that assumed under the assumption that the thick invert drainage blanket and collector drains should continue to function.

Analytical Closed-Form Solution

An analysis of the interaction between a liner and the surrounding rock mass needs to be carried out to evaluate the rate of leakage and the hydraulic head drop across the liner. Fernandez (1994) presented a hydraulic model for the analysis of the hydraulic interaction between the lining and the surrounding ground.

When a tunnel is unlined, the hydrostatic water pressure is exerted directly on the tunnel boundary. When a liner is placed, the total head loss across the liner-rock system, Δh_w, is composed of head losses across the liner, Δh_L, head losses across the grout zone if any, Δh_G, and head losses across the medium, Δh_m. The head loss across the liner is systemically presented in figure.

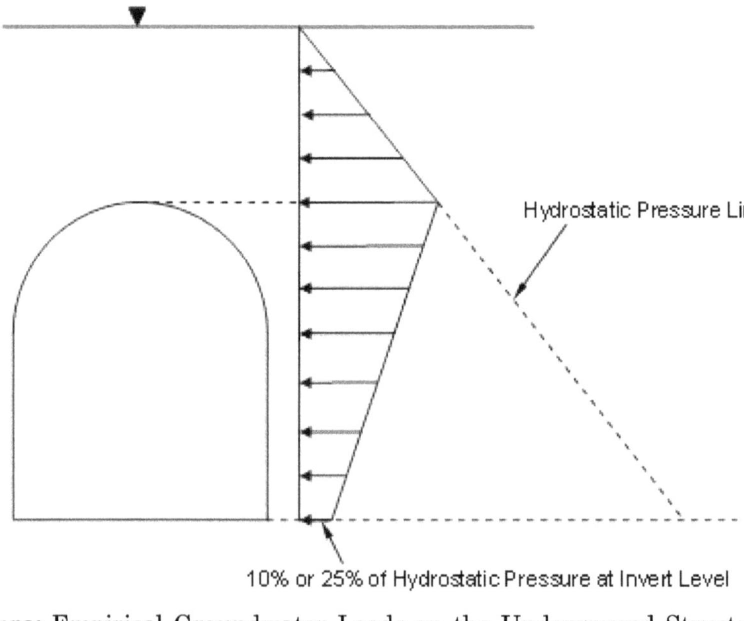

Figure: Empirical Groundwater Loads on the Underground Structures

Fernandez (1994) indicated that the head loss across the liner normalized by the total head loss across the system is expressed by:

$$\frac{\Delta h_l}{\Delta h_W} = \frac{1}{1 + C\dfrac{K_L}{K_M}} \quad C = \frac{In(L/b)}{In(b/a_1)}$$

where k_L and k_m are the permeability of the liner and surrounding ground, respectively, and b and a_I are outside and inside radii of the lining, respectively. L can be estimated as twice the depth of the tunnel below the groundwater level unless a drainage gallery is excavated parallel to the tunnel. If a drainage gallery is drilled parallel to the tunnel, the value of L can be adjusted and set equal to the centre to centre distance between the pressure tunnel and the gallery. In common engineering practice, the hydraulic head loss across the liner could be 80-90 % of the net hydraulic head for relatively impermeable liners, with k_L/k_m approximately equal to 1/80 to 1/100.

Numerical Methods

A finite element seepage analyses can be used to predict hydraulic response of the ground in the vicinity of the tunnel construction. In the finite element analysis, both the tunnel liner and surrounding ground are idealized as isotropic and homogeneous media. The actual flow regime through the jointed rock mass and cracked concrete may be a fluid flow through the fracture networks; therefore, the absolute value of the hydraulic and mechanical response of the rock mass and concrete liner may differ from the prediction based on the assumption of isotropic, homogeneous, porous media.

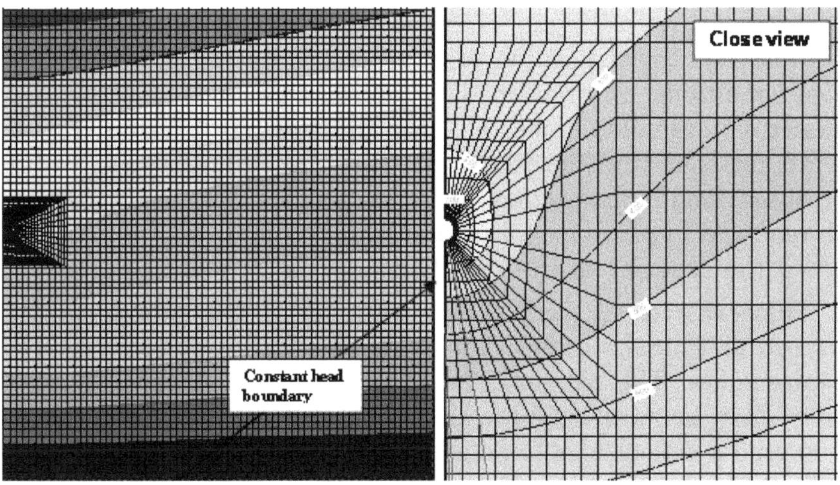

Figure: Two Dimensional Finite Element Groundwater Flow Model Analysis

It should be noted that this Finite Element Method analysis was focused on the global behaviour of the rockmass, treating the rockmass as a porous, continuum, isotropic rather than discrete (i.e., blocky) material. The finite element approach (i.e., rock mass rather than discrete rock blocks) considers the equivalent rockmass permeability, where the effects of hydraulic characteristics of fluid flow through rock joints are accounted for and approximated by the equivalent rockmass permeability. This approach has been used frequently for groundwater flow problems in the field of tunnel engineering. However, estimating the equivalent rock mass permeability closer than an order of magnitude is a great challenge and certainly requires special attention. Use of discrete element analysis is sometimes very difficult because it requires detailed input parameters of the rock joints such as joint attitudes, joint spacing, joint connectivity, hydraulic apertures of the joints, normal and shear stiffness.

Coupling effect of the mechanical and hydraulic behaviour of rock joints also requires understanding of the relationship between mechanical closure and hydraulic aperture of the joints. Without proper input parameters, the results from the discrete element analysis would not be reliable.

Drained Versus Undrained System

Drained Waterproofing System Drained waterproofing systems reduce hydrostatic loads on structures, enabling thinner and more lightly reinforced liners to be designed. In a fractured rock mass, high groundwater inflows often enter drained systems (even after rock mass grouting) resulting in increased pumping costs. High inflows can also increase the deposition of calcium precipitate in pipes. Under these conditions, an undrained system may be more efficient. In the drained waterproof systems, the pipes and drainage layers are required to remain open and flowing to prevent the build-up of hydrostatic pressures. Regular inspections and maintenance of the drainage system are required to prevent hydrostatic loads rising to a level that could exceed the capacity of the structure.

Allowable water infiltration rate varies depending upon the purpose of tunnel, tunnel dimension and local environmental law requirements. The rate of allowable infiltration acceptable to the owner shall be as specified in the contract documents. Some owners have used a rate of 1 gallon/minute per 1000 ft of tunnel length. The local infiltration limit is 0.25 gallon per day for 10 square feet of area, and 1 drip per minute at any location.

Undrained Waterproofing System Undrained waterproofing systems incorporate a membrane that extends around the entire tunnel perimetre with the aim of excluding groundwater completely. Final linings are designed for full hydrostatic water pressures. Thus, flat-slab walls or inverts are generally thick, whereas curved liners generally require less strength enhancement. The increased volume of excavation to create a curved or thicker invert is offset by reduced excavation for a gravel invert and sidewall pipes.

No groundwater drainage system is provided in the undrained system, resulting in cost savings from eliminating perforated sidewall pipes, porous concrete, transverse pipes and the gravel layer. If initial construction is of a high quality, operations and maintenance costs are low, because there is reduced pumping and without groundwater entering the tunnel drainage system, calcite deposits accumulate much more slowly. Reducing the inflow and drawdown also minimises the chance of having to deal with contaminated water.

Uplift Condition

The question of uplift forces on the tunnel lining must also be considered for tunnels in rock, especially if the tunnel is to be undrained. When squeezing or swelling conditions are encountered they act upwards on the invert just as they act anywhere else around the perimetre. When the tunnel is first driven, these forces may be at least partially relieved by the act of excavation and also somewhat reduced because gravity works in opposition to these upward forces. As time passes, however, the upward forces from swelling and/or squeezing come into full effect, that is, equal to those occurring anywhere else around the opening. Even if these rock loads should not be developed, the water head on an undrained tunnel will certainly be equal to the in-situ groundwater pressure.

Whether it is swelling, squeezing, water pressure or any combination thereof the invert of the tunnel will be subjected to upward forces. Typically, this means that the invert should be modified to a curved geometry to react to these upward forces – it is far easier to develop a stable curved structure than it is to permanently stabilise a flat invert. Even without squeezing or swelling the uplift water load can be quite expensive to resist with a flat invert as compared to a curved one. Thus, the most economical solution is usually to go directly to a curved configuration wherein the curved shape (when supported by steel ribs) will carry almost twice the load it will carry with a straight invert or straight sides (Proctor, 1968).

Waterproofing

For the most part tunnels in rock are waterproofed by a sandwich consisting of:

- A geotechnical drainage fabric that is put in place directly against the rock either continuously or in strips. This may be held in place by pins or nails driven or shot into the rock.

- Next a continuous waterproof membrane is installed. This membrane may be high density polyethylene (HDPE) or polyvinylchloride (PVC) or other similar material. To be continuous the membrane has to be cut and fit to all strange shapes and corners encountered and welded together (by heat) to make a continuous waterproofing membrane within the tunnel. Successful installation is quite dependent upon workmanship in three areas:
  - o Avoiding puncturing, tearing and the like of the membrane.
  - o Correctly making and testing all joints.
  - o Connecting the waterproof membrane to the wall without introducing leaks.
- Finally a cast in place lining of concrete is placed to hold the sandwich together and to provide the desired inner surface of the tunnel. Of course, the challenge is to get the concrete placed without damaging the membrane(s), this is especially challenging when the cast in place concrete must be reinforced.

Unlined and partial lined tunnels are common in many short mountainous tunnels in competent rock and stabilised with or without patterned rock bolts on the exposed rock. Groundwater inflows are tolerated and collected.

Chapter 5

Soft Ground Tunnelling

Human kind has been excavating in soft ground for thousands of years. Archeological digs in Europe and elsewhere show that all kinds of tools were used by our ancestors to excavate soil (mostly for "caves" in which to live): bones, antlers, sticks, rocks and the like. However, there are tunnels in Europe that were built by the Romans, are over 2000 years old, and are still in service carrying water. As the population grows and we demand more and more transportation services, there can be no doubt that the requirement for tunnels will also grow. Through it all, the art of tunnel design and construction will also continue to develop, but it is doubtful that this art will ever develop into a science comparable to structural design.

The structural engineer can specify both the configuration and the properties in great detail; the tunnel engineer must work with existing materials that cannot be specified and, in addition, are constantly changing, often dramatically. Problematic soft ground conditions such as running sand and very soft clays. Mining sequentially through soft ground based on the sequential excavation method (SEM) principles. The data needed for analysis and design.

Influence of the Support System on Equilibrium Conditions

Most tunnel openings are supported at some stage of construction. The behaviour of a tunnel opening and a support system is dependent on the time and manner of the placement of the support and its deformational characteristics.

The reasons for providing support are manifold. Sometimes support is required for the immediate stability of the opening. It may be furnished even before excavation, for example by air pressure, forepoling or ground improvements. Under these circumstances the interaction between the medium and thc supporting agent commences during or before excavation. When a shield is used for immediate support, a lining is erected inside the shield, and the annular void cleared by the shove of the shield is at least partly filled with pea-gravel and/or grout. The lining may be intended as a permanent support consisting, for example, of precast concrete segments. It may alternatively be a relatively flexible one in which a stiffer permanent lining will later be constructed. In this event, at least three different equilibrium conditions must be considered.

Where there is need for long-term but not immediate support, the support may be constructed at some distance behind the face. A partial relaxation with associated movements may then take place before the support interacts with the medium. Often a liner is erected and expanded into contact with the medium. The expansion induces a prestress in both the liner and the medium and influences subsequent deformations.

Even where instability or collapse of the opening is not imminent, support may still be required for various reasons, usually to control or limit deformations. Large deformations may lead to undesired settlements of the ground surface or to interference with other structures. Such deformation must be restrained at a suitably early stage. Deformations of a soil or rock mass commonly result in an undesirable reduction in strength and coherence of the medium. In a jointed or weak rock the material above the opening tends to loosen and may sooner or later exert considerable loads on the support. These loads are reduced if loosening is prevented by suitable support.

Although the initial stability may be satisfactory, conditions may be such that final equilibrium cannot be reached without support. This may occur in jointed rock mass subject to progressive loosening, in creeping or swelling materials, and in materials whose strength decreases with time. Except in such creeping materials as salts, these long term phenomena are associated with volume changes.

It is impossible and undesirable to avoid deformations in the soft ground altogether. Some movement is necessary to obtain a favourable distribution of loading between the medium and the support system. In each instance, the engineer must determine how much movement is beneficial to the behaviour of the tunnel, and at what movements the effects will become detrimental. The engineer's conclusions regarding these matters determine whether and where restraints are to be applied to the tunnel walls. His conclusions also determine the character and magnitude of those restraints. In tunnels in hard rock the beneficial movements take place almost immediately, and subsequent movements are likely to lead to loosening and additional loading. Hence, in this case rapid construction of supports is usually desirable.

It is apparent that many factors determine whether and where a support system should be constructed for structural reasons alone. The final choice of whether and where supports are actually employed is influenced by additional factors such as the psychological well-being of the workers, or the economy that might be achieved by adopting a uniform construction procedure throughout the same tunnel even though the properties of the medium vary.

No matter what the reason for using restraints, the loads to which a support will be subjected depend on the stage of equilibrium prevailing at the time the support is introduced. Thus, if final equilibrium has been reached before support is provided, the support may not receive loads from the medium at all. On the other hand, when support is furnished before final equilibrium has been established, new boundary conditions are superimposed on the conditions existing at the time the support is constructed. The new final conditions depend on the time the support was provided and involve the interaction between the support and the medium. If a stiff support could be installed in the medium before excavation by an imaginary process that did not in any way disturb the remaining material, it would be subjected to stresses resembling those of the in-situ condition existing before the excavation. However, the at least temporary reduction of the radial stresses to atmospheric pressure (or to the air pressure in the tunnel), as well as many other activities, generally introduce such deformations into the medium that the stresses ultimately acting on the tunnel support bear little or no resemblance to the initial stresses in the medium.

Procedures for the analysis and design of tunnel supports are necessarily simplified, but they should be based on the considerations of equilibrium and deformations briefly outlined above. In addition, a number of factors which are not directly related to the interaction between a support system and the medium are significant in the actual design of supports. Such factors, which are dealt with in the following section, sometimes even override considerations of structural interaction." (After Deere, 1969).

Excavation Methods

Shield Tunnelling: Generally soft ground tunnelling did not become viable until the introduction of the tunnel shield (accredited to Sir Marc Brunel), except for small hand-excavated openings in soft ground and somewhat larger ones in soft rock, tunnelling. Brunel wrote: "The great desideratum (sic) therefore consists in finding efficacious means of opening the ground in such a manner that no more earth shall be misplaced than is to be filled by the shell or body of the tunnel and that the work shall be effected with certainty" (Copperthwaite, 1906). In other words, never open more than is needed, can be excavated rapidly, and quickly supported. Brunel patented a circular shield in 1818 that was described by Copperthwaite (1906) as covering "every subsequent development in the construction and working of tunnel shields."

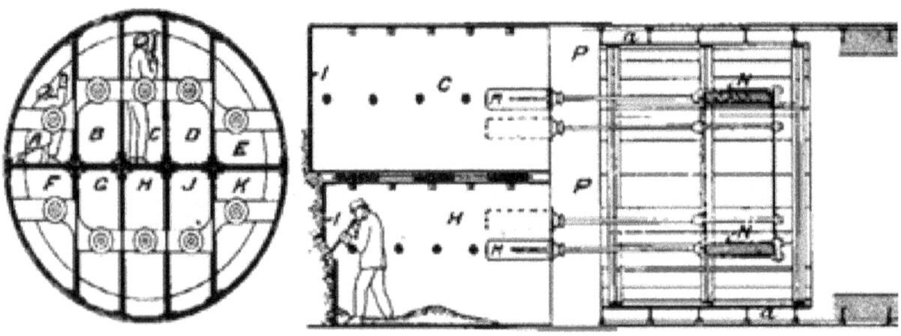

A-K Work Cells

N Reaction Framing

1 Breasting-Boards

P Shield Advance

M Hydraulic Jack

a Tunnel Lining

Figure: Patent Drawing for Brunel's Shield, 1818 (Cooperthwaite, 1906).

If we fast forward, we find that nearly all soft/ground tunnels driven in North America into the 1960's and the early 1970's were mostly under 10 feet (3 m) diametre and driven using the basic concepts of the Brunel tunnel shield; viz, compartmentalized, face breasting with timber and lots of hand labour.

In ground conditions that required a higher level of support than the basic Brunel shield, compressed air was commonly used (actually from the mid 1800's into the 1980's). When used correctly, compressed air provided the needed support and allowed many tunnels to be completed that would otherwise not have been possible. Because of the decompression required and all the associated equipment and procedures, not to mention the potential hazards to the workers, e.g., the bends or even death, compressed air has largely been eliminated as a tunnelling adjunct. Starting in the late 1960's and early 1970's, mechanization began to be introduced by incorporating excavating machines within the circular shields, hence the term digger shield.

Figure: Digger Shield with Hydraulically Operated Breasting Plates on Periphery of Top Heading of Shield used to Construct Transit Tunnel.

However, digger shield machines too often met with poor results and were usually unsatisfactory for three reasons:

1. Ground loss occurred ahead and above the shield when retracting the doors or poling plates. Typically, the orange-peel doors could not be retracted in tune with the forward progress of the shield. Also when retracting the doors, the miner does not have access to deal with running ground. Thus the machine encouraged unwanted ground movement, rather than controlling it.
2. Maintaining the right soil "plug" in the invert was always a headache.
3. Mounting the digger in the centre created a "Catch 22": if the ground movement in the centre became excessive, the only way to stop it was to cram the digger bucket into the face. However, that made it impossible to excavate and move the shield forward because to do so meant the bucket had to be moved, allowing the face to fail.

Shields with open faced wheeled excavators were another, early step in mechanization of soft-ground machines that have some things in common with their cousins the hard rock TBMs. Wheeled excavators were used with success in firm ground conditions, but not so well in running or fast raveling ground conditions. In some ground conditions this arrangement was marginally successful but in general it was not possible always to control the amount of ground allowed through the wheel to be equal to only that described by the cutting edge of the shield. As a turning point in global tunnelling equipment development, soft ground tunnel shields equipped with wheeled excavators were exported to Japan . Further development of soft-ground tunnelling machines was flat in the USA for many years, Japan, however, took a good idea, invested heavily in equipment development and within a decade or so exported vastly improved tunnelling methods back to the USA in the form of pressurized-face tunnelling machines.

Thus as soft ground tunnelling in the USA was affixed with traditional shield tunnelling, the Japanese, Europeans (read that Germans), the UK, and Canadians were developing two more "modern" machines - the earth pressure balance machine (EPB) and the slurry face machine (SFM).

At first-hand, these machines are similar in that they both have:

1. A revolving cutter wheel.
2. An internal bulkhead that traps cut soil against the face (hence, they are called closed face), and that maintains the combined effective soil and water pressure and thereby stabilises the face.
3. No workers are at the face but rely on mechanization and computerization to control all functions, except segment erection (to date).
4. Precast concrete segments erected in the shield tail, with the machine advanced by shoving off those segments.

The actual functioning of the machines, however, has some distinct differences: in the EPB the pressure is transmitted to the face mechanically, through the soil grains, and is reduced by means of friction over the length of the screw conveyor. Control is obtained by matching the volume of soil displaced by forward motion of the shield with the volume of soil removed from the pressurized face by that screw conveyor and deposited (at ambient pressure) on the conveyor or muck car. Clearly the range of natural geologic conditions that will result in suitably plastic material to transfer the earth pressure to the face and, at the same time, suitably frictional to form the "sand plug" in the screw conveyor is rather limited - generally only combinations of fine sands and silts.

Figure: Earth Pressure Balance Tunnel Boring Machine (EPB) (Lovat).

In contrast, the SFM transmits pressure to the face hydraulically through a viscous fluid-formed by the material cut and trapped at the face and mixed with slurry (basically bentonite and water). In this case the pressure transmitted can be controlled by means of pressure gages and control valves in a piping system. By this system a much more precise and more consistent pressure control is attained. The undesirable aspect of this system is the separation plant that has to be built and operated at the surface to separate the slurry from the soil cuttings for disposal and permit re-use of the slurry. Finding a site for the slurry separation that is satisfactory for the process and acceptable to the public can present interesting challenges.

Figure: Slurry Face Tunnel Boring Machine (SFM)

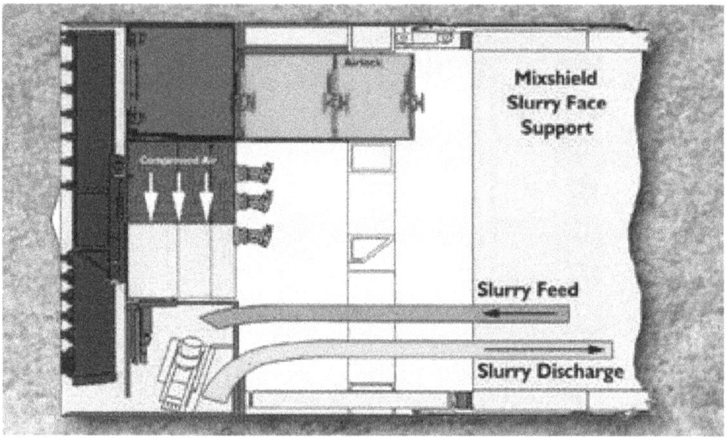

Figure: Simplified Cross-section of Slurry Face Tunnel Boring Machine (SFM)

During the last decade or so, great strides have been made in developing new families of conditioning agents that can be used in both types of closed face machines. These additives tend to blur the distinctions portrayed above and widen the range of applicability of both types of machines. Indeed, we predict that in another decade we will not be talking about the two type of machines but rather a new family of machines that will operate interchangeably and with equal efficiency as an open face wheel machine in stable ground or as a closed face machine (with conditioners) that will cut any type of soft ground. Herrenknecht, for one, is already moving ahead with development of this new breed of machine.

Throughout all of this development, the role of the miner at the tunnel face is steadily being diminished. With any closed face machine, the miner is not doing any excavating or breasting of the face. The miner is operating machines that, unfortunately, can not always do the job as advertised. (After Hansmire and Monsees, 2005)

Choosing between Earth Pressure Balance Machines and Slurry Tunnelling Machines

The choice of the type of closed-face tunnelling machine and its facilities is a critical decision on a soft-ground tunnelling project. This decision will be guided by thorough assessment of the ground types and conditions to be encountered and by numerous other aspects.

Other aspects that will influence the choice include the particular experience of the project's contractor, the logistics and configuration of the works, and requirements of the contract as a means to ensure that the client's minimum specification is met. The initial choice is guided by reference to the grading envelope of the soils to be excavated. Since it is likely that the geology will fall into more than one envelope, the final choice may require a degree of compromise or development of a dual-mode open/closed-faced TBM system or a dual slurry/EPB system. Review of Ground Types In many tunnel drives the conditions encountered along the route may vary significantly with a resulting need to specify a system capable of handling the full range of expected conditions. Closed-face tunnelling machines can be designed and manufactured to cope with a range of ground conditions. Some machines are capable of handling many or all of this range of anticipated conditions with a limited degree or reconfiguration for efficient operation.

There have been several attempts to classify the naturally occurring range of soft-ground characteristics from the tunneller's perspective. This work was summarised most recently by Whittaker and Frith (1990) and the following categorization is based partly on their work. It consists of eight categories of physical ground behaviour that may be observed within the soft-ground tunnel excavation range. The characteristics are summarised in Table 7-5 . Each of these may be associated with particular types of soils.

Selection Criteria Based on Particle Size Distribution and Plasticity An SFM is ideal in loose waterbearing granular soils that are easily separated at the separation plant. By contrast SFMs have problems dealing with clays and some silts.

If the amount of fines (particles smaller than 60 mm or able to pass through a 200 sieve) is greater than 20% then the use of an SFM becomes questionable although it is not ruled out. In this situation it will be the difficulty in separating excavated spoil from the slurry, rather than the operation of the TBM, that is likely to affect critically the contract program and the operating cost.

An EPBM will perform better where the ground is silty and has a high percentage of fines both of which will assist the formation of a plug in the screw conveyor and will control groundwater inflows. A fines content of below 10% may be unfavourable for application of EPBMs. For an EPBM the costs of dealing with poorly graded or no-fines soil will be in the greater use of conditioners and possibly, in extreme cases, the use of positive displacement devices, such as rotary feeders or piston dischargers, at the screw conveyor discharge point to maintain EPB pressures.

Higher plasticity index (PI) clays ('sticky clays') can lead to 'balling' problems and increased problems at the separation plant for SFMs.

Similarly these materials can be problematic for EPBMs where special attention is required in selecting the most appropriate conditioning agents.

Table 1: Soft-ground Characteristics

Ground	*Description*
Firm ground	Ground in which the tunnel can be advanced safely without providing direct support to the face during the normal excavation cycle and in which ground support or the lining can be installed before problematic ground movement occurs. Where this short-term stability may be attributable to the development of negative pore pressure in the fine grained soils, significant soil movements and/or ground loading of the tunnel lining may occur later. Examples may include stiff clays and some dewatered sands. A closed-face tunnelling machine may not be needed in this ground type.
Raveling ground	Ground characterised by material that tends to deteriorate with time through a process of individual particles or blocks of ground falling from the excavation surface. Examples may include glacial tills, sands and gravels. In this ground a closed-face tunnelling system may be required to provide immediate support to the ground.
Running or flowing ground	Ground characterised by material such as sands, silts and gravels in the presence of water, and some highly sensitive clays that tend to flow into an excavation. Above the water table running ground may occur in granular materials such as dry sands and gravels. Below the water table a fluidized mixture of soil and water may flow as a liquid. This is referred to as running or flowing ground. Such materials can sometimes pass rapidly through small openings and may completely fill a heading in a short period of time. In all running or flowing ground types there will be considerable potential for rapid over-excavation. Hence, a closed-face tunnelling system will be required to support such ground safely unless some other method of stabilisation is used.
Squeezing ground	Ground in which the excavation-induced stress relief leads to ductile, plastic yield of ground into the tunnel opening. The phenomenon usually is

exhibited in soft clays and stiffer clays over a more extended period of time. A closed-face machine may be required to provide resistance to squeezing ground, although in some conditions there is also a risk of the TBM shield becoming trapped.

Swelling ground — Soil characterised by a tendency to increase in volume due to absorption of water. This behaviour is most likely to occur either in highly over-consolidated clay or in clays containing minerals naturally prone to significant swelling. A closed-face machine may be useful in providing resistance to swelling ground although, as with squeezing ground, there is a risk of the shield becoming trapped.

Weak rock — Weak rock may be regarded effectively as a soft-ground environment for tunnelling because systems used to excavate soft-ground types may also be applied to weak rock materials such as chalk. Weak rock will often tend to be self-supporting in the short term with the result that closed face tunnelling systems may not be needed. However, groundwater may be significant issue. In these instances a closed-face machine is an effective method of protecting the works against high volumes of water ingress that could also be under high hydrostatic pressure.

Hard rock — A closed-face TBM may also be deployed in normally self-supporting hard rock conditions. The main reason would be to provide protection against groundwater pressures and prevent inundation of the heading.

Mixed ground conditions — Potentially, the most difficult of situations for a closed-face tunnelling system is that of having to cope with a mixture of different ground types either along the tunnel from zone to zone or sometimes from metre to metre, or within the same tunnel face. Ideally the vertical alignment would be optimized to avoid, as far as possible, a mixed ground situation, however, in urban locations the alignment may be constrained by other considerations. For changes in ground types longitudinally, a closed-face machine may have to convert from a closed-face pressurized mode to an open non-pressurized mode when working in harder ground types to avoid over stressing the machine's mechanical functions. Such a change may require some modification of the machine and the reverse once again when the alignment enters a reach of soft, potentially unstable ground. In the case of mixed ground types across the same face, the tunnelling machine will almost certainly have to operate in a compromise configuration. In such cases great care will be needed to ensure that this provides effective ground control. A common problem, for example, is a face with a hard material in the bottom and running ground at the top. In this situation the TBM will generally advance slowly while cutting the hard ground but may tend to draw in the less stable material at the top leading to over-excavation of the less stable material and subsequent subsidence or settlement at the surface. Different ground types at levels above the tunnel will also be of significance. For example, in the event that over-excavation occurs, the presence of running or flowing materials at horizons above the tunnel will increase the potential quantity of ground that may be over-excavated and again lead to subsidence or surface settlement. Another potential problem occurs when a more competent layer exists over potentially running ground in which case possible over-excavation would create voids above the tunnel and below the competent material, giving rise to potential longer-term instability problems.

Permeability As a general guide the point of selection between the two types of machines is a ground permeability of $1x10^{-5}$ m/s, by using SFMs applicable to ground of higher permeability and EPBMs for ground of lower permeability. However, an EPBM can be used at a permeability of greater than $1x10^{-5}$ m/s by using an increased percentage of conditioning agent in the plenum. The choice will take into account the content of fines and the ground permeability.

Hydrostatic Head High hydrostatic heads of groundwater pressure along the tunnel alignment add a significant concern to the choice of TBM. In situations where a high hydrostatic head is combined with high permeability or fissures it maybe be difficult to form an adequate plug in the screw conveyor of an EPBM. Under such conditions an SFM may be the more appropriate choice especially as the bentonite slurry will aid in sealing the face during interventions under compressed air. Settlement Criteria Both types of machine are effective in controlling ground movement and surface settlement - providing they are operated correctly. While settlement control may not be overriding factor in the choice of TBM type, the costs associated with minimising settlement should be considered. For example, large quantities of conditioning agent may be needed to reduce the risk of over-excavation and control settlement if using EPBM in loose granular soils. Final Considerations Other aspects to consider when making the choice between the use of an SFM or an EPBM include the presence of gas, the presence of boulders, the torque and thrust required for each type of TBM and, lastly, the national experience with each method. These factors should be considered but would not necessarily dictate the choice.

The overriding decision must be made on which type of machine is best able to provide stability of the ground during excavation with all the correct operational controls in place and being used.

If both types of machine can provide optimum face stability, as is often the case, other factors, such as the diametre, length and alignment of the tunnel, the increased cutter wear associated with EPBM operation, the work site area and location, and spoil disposal regulations are taken into consideration. The correct choice of machine operated without the correct management and operating controls is as bad as choosing the wrong type of machine for the project.

Sequential Excavation Method (SEM)

In addition to shield tunnelling methods discussed above, soft ground tunnels can be excavated sequentially by small drifts and openings following the principles of the Sequential Excavation Method (SEM), aka New Austrian Tunnelling Method (NATM) first promulgated by Professor Rabcewicz (1965). The SEM has now been defined as "a method where the surrounding rock or soil formations of a tunnel or underground opening are integrated into an overall ring-like support structure and the following principles must be observed:

- The geotechnical behaviour must be taken into account
- Adverse states of stresses and deformations must be avoided by applying the appropriate means of support in due time.
- The completion of the invert gives the above mentioned ring-like structure the static properties of a tube.
- The support means can/should be optimized according to the admissible deformations.

- General control, geotechnical measurements and constant checks on the optimization of the pre-established support means must be performed. (From ILF, 2004)

The underlying principle of SEM is actually the same as that stated by Sir Marc Brunel almost two centuries ago: "The great desideratum therefore consists in finding efficacious means of opening the ground in such a manner that no more earth shall be displaced than is to be filled by the shell or body of the tunnel and that the work shall be effected with certainty". (Copperthwaite, 1906) In other words, never open more than is needed, can be excavated rapidly, and quickly supported.

As applied to soft ground tunnelling, SEM generally cannot compete with tunnelling machines for long running tunnels but often is a viable method for:

- Short tunnels
- Large openings such as stations
- Unusual shapes or complex structures such as intersections
- Enlargements

Soft Ground Tunnelling

Ground Loads and Ground-Support Interaction

The main objectives of tunnel support system are to (1) stabilise the tunnel heading, (2) minimise ground movements, and (3) permit the tunnel to operate over the design life. In general, the first two functions are provided by an initial support system, whereas the third function is preserved with a final lining. The loading on the support system and its required capacity is dependent on when and how it is installed and on the loadings that will occur after it is installed. If the final lining is installed after the tunnel has been stabilised by initial support, the final lining will undergo very little additional loadings such as contact grouting pressures, thermal stresses, groundwater pressure, and/or time dependent loading (creep).

Generally, two types of loading have been considered to generate analytical solutions in tunnelling in soil - overpressure loading and excavation loading. If a ground is assumed to be isolated and a pressure is applied to the upper surface, it is considered to be *overpressure loading*, where the support system is placed in the ground when it was unstressed and the lining and the ground is normally handled by applying lateral pressure to the ground.

Practically the support system is never placed in an unstressed ground, instead it is placed in the opening after the initial deformation has occurred, and before any additional deformation occurs and the additional deformation induces loading into the support system. This induced loading is called *excavation loading*.

The load developed on the support system (initial support and final lining) is a function of relative stiffness of the lining with respect to the soil (ground-lining interaction). Both analytical solutions and numerical methods have been commonly used by design engineers to evaluate the effect of the relative lining stiffness on the displacement, thrust, moments in the lining for various loading configurations. The available methods are summarised in this section.

Analytical Solutions for Ground-Support Interaction

Analytical solutions for ground-support interaction for a tunnel in soil are available in the literature. The solutions are based on two dimensional, plane strain, linear elasticity assumptions in which the lining is assumed to be placed deep and in contact with the ground (no gap), i.e., the solutions do not allow for a gap to occur between the support system and ground.

Early analytical solutions by Burns and Richard (1964), Dar and Bates (1974), and Hoeg (1968) were derived for the overpressure loading, while solutions by Morgan (1961), Muir Wood (1975), Curtis (1976), Rankin, Ghaboussi and Hendron (1978), and Einstein et. al. (1980) were for excavation loading. Solutions are available for the full slip and no slip conditions at the ground-lining interface. Appendix E present the available published analytical solutions in Table E-2, as well as the background (excerpt from FHWA Tunnel Design Guidelines published in 2004). Appendix E also presents a sample analysis is in Table E-3, for a 22ft diametre circular tunnel with 1.5 ft thick concrete lining. The tunnel is located at 105 ft deep from the ground surface to springline and groundwater table is located 10 ft below the ground surface. Details of input parameters are shown in Table E-3a. The calculated lining loads from various analytical solutions are presented in Table E-3b. The result of finite element analysis is shown in figures.

Numerical Methods

Application of the analytical solutions is restricted when the variation of stress magnitude is significant with depth from the tunnel crown to the invert, such that assumptions made in the analytical solutions are not valid. Then, numerical method can be used to simulate support-ground interactions.

Numerical modelling has been driven by a perceived need from the tunnelling industry in recent times. It has led to large, clumsy and complex numerical models. Properly performed numerical modelling will lead engineers to think about why they are building it - why build one model rather than another - and how the design can be improved and performed effectively.

Tunnelling Induced Settlement

Ground settlement is of greater concern for soft ground tunnels than for rock for two reasons:

- Settlements are nearly always greater for soft ground tunnels.
- Typically more facilities that might be negatively impacted by settlements exist near soft ground tunnels than near rock tunnels.

With modern means and methods, both the designer and the contractor are now better equipped to minimise settlements and, hence, their impact on other facilities.

Sources of Settlement

Although there are a large number of sources or causes of settlement, they can be conveniently lumped into two broad categories: those caused by ground water depression and those caused by lost ground.

Groundwater Depression Groundwater depression may be caused by intentional lowering of the water during construction or by the tunnel itself (or other construction) acting as a drain. When either of these occurs the effective stress in the ground increases. Basic soils mechanics can then be applied to estimate the resulting settlement. For tunnels in granular soil the settlement due to this increase in effective stress is usually reflected as an elastic phenomenon requiring knowledge of the low stress modulus of the ground and calculation of the change in effective stress. Unless the soil contains silt or very fine sand, this elastic settlement will typically represent the majority of the total but its absolute value will also be relatively small.

For fine grained soils, the situation is a bit more challenging but certainly manageable using normal soil mechanics approaches. With fine-grained soils, the conditions are reversed. In most instances, the settlement is mostly due to consolidation brought on by the changes in effective stress and hence is analysed by the usual soil mechanics consolidation theories. In some instances, primarily if lenses of sands are contained in the soil, there may also be a relatively small contribution by elastic compression. In comparison to the settlement of granular soils, consolidation can lead to several inches of settlement when the consolidating soils are thick and the change in effective stress is significant.

Lost Ground Lost ground has a number of root causes (at least nine) and is usually responsible for the settlements that make the headlines. By definition, lost ground refers to the act of taking (or losing) more ground into the tunnelling operation than is represented by the volume of the tunnel. Thus it is highly reflective of construction means and methods. As will be discussed, modern machines can be a great help in controlling lost ground but in the end it usually comes down to quality of workmanship.

For the purposes of this manual, the causes of lost ground are lumped into three groups: face losses, shield losses and tail losses.

- Face losses results from movement in front of and into the shield. This includes running, flowing, caving, and/or squeezing behaviour of the ground itself or simply mining more ground than displaced by the tunnelling machine.
- Shield losses occur between the cutting edge and the tail of the shield. All shields employ some degree of overcut so that they can be maneuvered. In addition, any time a shield is off alignment, the shield yaws, pitches, or plows when brought back to alignment. Mother Nature abhors a vacuum and the surrounding soils begin to fill these planned or produced voids the instant they are produced. Note that a one inch overcut plus one-eighth inch hard facing on a 20 foot shield produces lost ground of nearly two percent if not properly filled [$1.125/12\ (20)\ 3.1416\ ]/\ (10)^2\ 3.1416 = 1.88\%$).
- Tail losses are similar to shield losses in that they are caused by the space being vacated by the tail itself as well as the extra space that must be provided between the tail and the support elements so those elements can be erected and so that they don't become "iron bound" and seize the tail shield. However, like the shield losses, these tail voids will rapidly fill with soil if they are not first eliminated by grouting and/or expansion of the tunnel support elements.

Settlement Calculations

Estimates of settlement in soft ground tunnelling are just that, estimates. The vagaries of nature and of construction are such that settlements cannot be estimated in soft ground tunnels to the same level of confidence as, say, the settlement of a loaded beam. In tunnelling we rely heavily on our experience with some assistance from analysis. Thus, there are two related methods to attack the problem: experience and empirical data.

Experience can be used where a history of tunnelling and of taking measurements exists. An example of this is Washington, D.C., where soft ground tunnels have been constructed in well-defined geology for over 40 years. During that time the industry has progressed from basic Brunel shields to the most current closed-face tunnelling machines. For this case it would be anticipated that an experienced contractor would achieve between 0.5 and 1.0 percent ground loss. An inexperienced contractor would attain 1.0 to 2.0 percent loss.

Table 2: Relationship between Volumes Loss and Construction Practice and Ground Conditions

Case	V_L *(%)*
Good practice in firm ground; tight control of face pressure within closed face machine in slowly raveling or squeezing ground	0.5
Usual practice with closed face machine in slowly raveling or squeezing ground	1.0
Poor practice with closed face in raveling ground	2
Poor practice with closed face machine in poor (fast raveling) ground	3
Poor practice with little face control in running ground	4.0 or more

When there is no record to rely upon, the design would have to be based strictly on empirical data and an engineering assessment of what the contractor could be expected to achieve with no track record to rely upon. In that case the above evaluations might be bumped up one-half percentage point each as an insurance measure

State-of-the-art pressurized-face tunnel boring machines (TBM) such as EPB and SFM minimise the magnitude of ground losses. These machines control face stability by applying active pressure to the tunnel face, minimising the amount of overcut, and utilising automatic tail void grouting to reduce shield losses. Typically, ground loss during soft ground tunnel excavation using this technology limits ground loss to 1.0 percent or less assuming excellent tunnelling practice (adequate pressure applied to the face and effective and timely tail void grouting).

The volume of ground loss experienced during tunnelling can be related to the volume of settlement expected at the ground surface (Peck, 1969). For a single tunnel in soft ground conditions, it is typically assumed the volume of surface settlement is equal to the volume of lost ground. However, the relationship between volume of lost ground and volume of surface settlement is complex. Volume change due to bulking or compression is typically not estimated or included in the calculations. Ground loss will produce a settlement trough at the ground surface where it can potentially impact the settlement behaviour of any overlying or adjacent bridge foundations, building structures, or buried utilities transverse or parallel to the alignment of the proposed tunnel excavation. Empirical data suggests the shape of the settlement trough typically approximates the shape of an inverse Gaussian curve.

The shape and magnitude of the settlement trough is a function of excavation techniques, tunnel depth, tunnel diametre, and soil conditions. In the case of parallel adjacent tunnels, surface settlement is generally assumed to be additive. The shape of the curve can be expressed by the following mathematical relationships (Schmidt, 1974).

$$w = w_{\max} \exp\left(\frac{-X^2}{2i^2}\right)$$

Where:

w = Settlement, x is distance from tunnel or pipeline centreline

i = Distance to point of inflection on the settlement trough

The settlement trough distance, i is defined as:

$i = KZ_{\bar{I}}$

Where:

K = Settlement trough parameter (function of soil type)

$Z_{\bar{I}}$ = The depth from ground surface to tunnel springline

The maximum settlement, w_{max} is defined as:

$$W\max = \frac{V_L \pi \left(\frac{D}{2}\right)^2}{2.5i}$$

Where:

V_L = Volume of ground loss during excavation of tunnel

D = A diametre of tunnel.

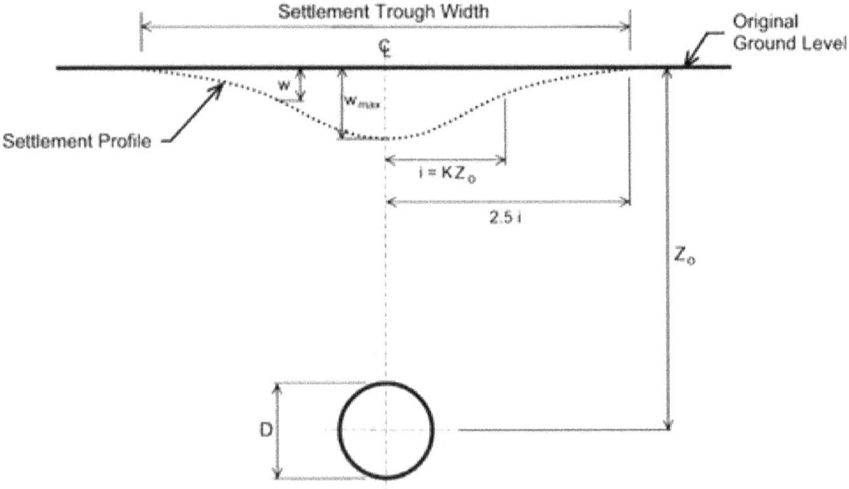

Figure: Typical Settlement Profile for a Soft Ground Tunnelling

For geometrics other than a single tunnel, adjustments of the types given below should be made to obtain settlement estimates:

- For *parallel tunnels* three or more diametres apart (centre to centre), surface settlements are usually reasonably well predicted by adding the individual bell curves of the two tunnels. In good ground and with good practice, this will often give workable approximations up to the point where the tunnels are two diametres apart. On the other extreme, when the tunnels are less than one and one-half diametres apart, the volume of lost ground assumed for the second tunnel should be increased approximately one level in severity before the bell curves are added. Intermediate conditions may be estimated by interpolation.
- For *over-and-under tunnels*, it is usually recommended that the lower tunnel be driven first so that it does not undermine the upper tunnel. However, driving the lower tunnel will disturb the ground conditions for the upper. This effect may be approximated by increasing the lost ground severity of the second (upper) tunnel by approximately one level before adding

the resulting two settlement estimates to approximate the total at the surface.

The width of the settlement trough is measured by an i value, which is theoretically the horizontal distance from the location of maximum settlement to the point of inflection of the settlement curve. The maximum value of the surface settlement is theoretically equal to the volume of surface settlement divided by 2.5 i.

The ground settlement also can be predicted by numerical methods. The numerical method is extremely useful when the tunnel geometry is not a circular or horse-shoe shape since analytical/empirical method is not directly applicable.

Impact on and Protection of Surface Facilities

Evaluation of Structure Tolerance to Settlement

Evaluation of structural tolerance to settlement requires definition of the possible damage that a structure might experience. Boscardin and Cording (1989) introduced three damage definitions for surface structures due to tunnelling induced settlement:

1. *Architectural Damage:* Damage affecting the appearance but not the function of structures, usually related to cracks or separations in panel walls, floors, and finishes. Cracks in plaster walls greater than 1/64-in. wide and cracks in masonry or rough concrete walls greater than 1/32-in. wide are representative of a threshold where damage is noticed and reported by building occupants.
2. *Functional Damage:* Damage affecting the use of the structure, or safety to its occupants, usually related to jammed doors and windows, cracking and falling plaster, tilting of walls and floors, and other damage that would require nonstructural repair to return the building to its full service capacity.
3. *Structural Damage:* Damage affecting the stability of the structure, usually related to cracks or distortions in primary support elements such as beams, columns, and load-bearing walls.

A number of methods for evaluating the impact of settlements on building or other facilities have been proposed and used. In 1981, Wahls collected and studied data from other investigators (e.g., Skimpton and MacDonald, 1956: Grant, Christian, and Vanmarked; Polshin and Tokar) plus his own observations (totalling more than 193 cases). From that study Wahls proposed the correlation of angular distortion (the relative settlement between columns or measurement points) and building damage.

As an alternative initial screening method, Rankin (1988) proposed a damage risk assessment chart based on maximum building slope and settlement.

Table 3: Limiting Angular Distortion

Category of Potential Damage	*Angular Distortion*
Danger to machinery sensitive to settlement	1/750
Danger to frames with diagonals	1/600
Safe limit for no cracking of building	1/500
First cracking of panel walls	1/300
Difficulties with overhead cranes	1/300
Tilting of high rigid building becomes visible	1/250
Considerable cracking of panel and brick walls	1/150
Danger of structural damage to general building	1/150
Safe limit for flexible brick walls*	1/150

[a] Safe limit includes a factor of safety.

Table 4: Damage Risk Assessment Chart

Risk Category	*Maximum slope of building*	*Maximum settlement of building (mm)*	*Description of risk*
1	Less then 1/500	Less than 10	Negligible; superficial damage unlikely
2	1/500 - 1/200	10-50	Slight; possible superficial damage which is unlikely to have structural significance
3	1/200 - 1/50	5-075	Moderate; expected superficial damage and possible structural damage to bulidings, possible damage to relatively rigid pipelines
4	Greater than 1/50	Greater than 75	High; expected structural damage to buildings. Expected damage to rigid pipelines, possible damage to other pipelines

Mitigating Settlement

Where the settlement is or would be caused by groundwater lowering the first, and usually the simplest, approach is simply to reduce or eliminate the conditions causing or allowing dewatering. This could include, for example:

- Reduce drawdown at critical structures by reinjecting water, using impervious cutoff walls and the like.

- Using closed, pressurized face tunnelling machines so that drawdown can not occur. Pressure at the face should be equal to the groundwater head.
- Grouting the ground around the tunnel to eliminate water inflow into the tunnel.

Where the settlement is or could be caused by lost ground in the tunnelling operation that settlement can nearly always be mitigated with proper construction means and methods. For example consider:

- Requiring a closed face, pressurized TBM (EPB or SFM) and keep the pressure at least equal to if not greater than the combined soil and groundwater pressure in the ground at tunnel level.
- Immediately and completely grout the annular space between the tunnel lining and the ground at the tail of the machine. Use automated grouting systems that will not permit the machine to advance without this void being simultaneously grouted.
- Control the operation (steering) of the machine so that it is not forced to pitch or yaw to make excessive alignment corrections. Each one percent of correction translates to a potential 1.5 percent of ground loss.
- Use compaction or compensation grouting to "make up" for ground loss before it migrates to the building.
- Treat areas of loose soils by consolidation or jet grouting before tunnelling into them.

Structure Protection

The concept of and methods for structure protection are already woven into earlier paragraphs. First and foremost are the tunnelling procedures of maintaining face pressure (control) and immediately grouting to fill the annular (or any other) void.

The next step is ground improvement either by consolidation or jet grouting and, closely related compensation or compaction grouting.

As a last resort, to be applied when all else appears to be unsuccessful and or unworkable, is underpinning. Like the use of compressed air, this method is now seldom used because modern tunnelling techniques make it unnecessary. At times, as with the Pershing Square garage in Los Angeles it is still applicable, but most of the time practitioners believe it to have the possibility to do more damage than to be beneficial. Typical steps of underpinning method are summarised as follow:

- Break out and hand excavate down to (or nearly to) the potentially impacted foundation.
- Install piles or other founding elements to a bearing below and/or outside the impacted foundation and the tunnel.
- Install a needle beam or similar method to transfer the impacted foundation load to the new elements.
- Preload the new elements, i.e., unload the impacted foundation onto those new elements
- Cut or release any load to the impacted foundation. At this point all load is transferred through the new elements to a bearing location/condition that is completely independent of the tunnelling operation and the tunnel.
- As required or necessary remove or leave in place the original foundation.

Soil Stabilisation and Improvement

Purpose

Until fairly recently essentially all the design effort for tunnels in soft ground was to provide a support system or systems that would stabilise the existing ground during construction and then, perhaps with some modification, would permanently support the ground and provide an opening suitable for the long term mission of the tunnel. In the last two or three decades, however, the situation has changed such that in some applications a dual approach is taken. First, the characteristics of the ground are modified by stabilisation and/or improvement to make that ground contribute more to its own stability. Then, secondly a supplementary but less costly support/lining system is installed to make the tunnel perform for its full lifetime. In this section the various methods of soil stabilisation and improvement are summarised. References with more details on these methods are also given.

Typical Applications

The decision to use soil stabilisation or improvement must be made on each individual case. This decision may sometimes be easy with there being no other way to construct the tunnel. More often, the decision comes down to a trade off among treating the ground, using high-tech machines, and/or a combination of the two. With all of the possibilities it can be said that there are now no unacceptable construction sites.

Table 5: Ground Treatment Methods

Challenging Ground Conditions	*Treatment Method(s)*
Weak Soils	• Vibro Compaction
	• Dynamic Compaction
	• Compaction Grouting
	• Permeation Grouting
	• Jet Grouting
Ground Water	• Dewatering
	• Freezing
	• Grouting
Unstable Face	• Soil Nails
	• Spiling
	• Soil Doweling
	• Micro Piles
Soil Movement	• Compensation Grouting
	• Compaction Grouting

It is to be noted that the boundaries between both ground conditions and treatment methods are not fixed. Also, the use of vibrocompaction techniques or dynamic compaction is typically applicable at or near the tunnel portals as these techniques are applied to the ground surface and are not effective beyond about 100 ft depth for vibro compaction and 35 ft depth for dynamic compaction. Both are generally effective only in granular soils.

Readers are referred to the Ground Improvement Methods Reference Manual (FHWA, 2004) for more detailed discussion for the soil stabilisation and improvement techniques presented below.

Reinforcement Methods

Soil Nails Soil nails may be used to stabilise a tunnel face in soil during construction. Steel or fiberglass rods or nails are installed in the face and the resulting reinforced block(s) are analysed for stability much as for usual slope stability analyses. Several methods (e.g., Davis, Modified Davis, German, French, Kinematrical, Golder, and Caltrans) are used for these analyses. Walkinshaw (1992) has studied these methods and concluded that all had some level of inconsistencies, such as:

- Improper cancellation of interslice forces (Davis method)
- Lateral earth pressures inconsistent with nail force and facing pressure distribution (all)
- No redistribution of nail forces according to construction sequence and observed measurements (all except Golder)
- Complex treatment and impractical emphasis on nail stiffness (Kinematical) (after Walkinshaw, 1992: Xanthakos, 1994)

For more discussion readers are referred to GEC No.7 Soil Nail Walls (FHWA, 2003), which also recommends that the Caltrans SNAIL program be used because it will handle both nails and tiebacks. However, it must be recognised that application of that or any other program must be tempered with appropriate judgment, measurements and case history experience.

Soil Doweling Soil doweling entails the installation of larger reinforcement members than does nailing. These dowels act in tension like soil nails but are large enough in cross section that they also develop some shearing resistance where they pass through the sliding surfaces.

Micropiles

As they are applied to tunnelling, micropiles are essentially the same as soil dowels. These are typically drilled piles two to six inches in diametre that contain a large reinforcing bar centreed in the hole and the hole backfilled with concrete. As opposed to pin piles that are typically installed at the surface (and that act in compression), the pin piles placed in tunnels typically act in tension and shear across the sliding surfaces.

Soil nails, soil dowels, and pin piles are typically installed at the face of the tunnel to stabilise that face for construction. Thus, they are continually being installed and mined out of the face. For ease in this mining operation, fiberglass bars (rods) are typically used in these applications because they are much easier to mine out and cut. In contrast, spiling tends to look out around the perimetre of the tunnel, thus steel is more likely to be used for spiling bars or plates.

Grouting Methods

All grouting involves the drilling of holes into the ground, the insertion of grout pipes in the holes, and the injection of pressurized grout into the ground from those pipes. The details of the operations, however, are distinctly different.

Permeation Grouting Permeation grouting involves the filling of pore spaces between soil grains (perhaps displacing water). The grout may be one of a number of chemicals (but is usually sodium silicate or polyurethane) or neat cement using regular, micro- or ultra- fine cement, along with chemicals and other additives. Once injected into the pore spaces, the grout sets and converts the soil into a stable, weak sandstone material. Permeation grouting usually involves grout holes at three to four feet centres with enough secondary holes at split spacing to verify that all the ground is grouted. If necessary to get full coverage all of the split spacing holes may have to be grouted and verification performed by the tertiary holes.

Compaction Grouting Compaction grouting uses a stiffer grout than does permeation grouting. In compaction grouting the goal is to form a series of grout bulbs or zones four to six feet above and around the tunnel crown. By pumping the stiff grout in under pressure these bulbs compress (densify) the ground above the tunnel and between the tunnel and overlying facilities.

The pipes for compaction grouting are pre-positioned and drilled into place and all the grouting pumps, hoses, header pipes, instrumentation and the like are in place before the tunnel drive begins. Instrumentation is read as the tunnel approaches and passes a facility and the grouting operation is adjusted real time in response to the movement readings. Actually, in most applications it is possible to either pre-heave the ground or to jack it back up (at least partially) by pumping more grout at higher pressures.

Compensation Grouting Compensation grouting is, in some ways, similar to compaction grouting. The goal is to monitor ground movements, primarily between the tunnel and any overlying facility. When it is apparent that ground is being lost in the tunnelling operation, a grout, typically slightly more liquid than the compaction grout mix, is injected to replace (compensate for) the lost ground. As indicated the differences between these two schemes are relatively minor - compaction grouting seeks to recompact the ground by forming grout bulbs, compensation grouting seeks to refill voids created by the tunnelling operations.

Jet Grouting Jet grouting is the newest of the grouting methods and is rapidly becoming the most widely used. Jet grouting uses high pressure jets to break up the soils and replace them with a mixture of excavated soils and cement, typically referred to as "soilcrete". There are a number of variations of jet grouting depending on the details of the application and on the experience and expertise of both the designer and the contractor.

Ground Freezing

As with much of tunnelling technology, ground freezing was developed first in the mining industry and was probably first used in sinking mine shafts. For a mine the shaft (and the mine) is located where the ore is. Thus, means of obtaining access in unfavourable ground conditions, of providing emergency support in unstable ground below the water table, and of maintaining stability of working faces below the water table, such as freezing, often had their roots in the mining industry.

In its simplest form, ground freezing involves the extraction of heat from the ground until the groundwater is frozen. Thus converting the groundwater into a cementing agent and the ground into a "frozen sandstone". The heat is extracted by circulating a cooling liquid, usually brine, in an array of pipes. Each pipe is actually two nested pipes, with the liquid flowing down the centre pipe and back out through the annulus between the pipes. When the pipes are close enough and the time long enough, the cylinders of frozen soil formed at each pipe eventually coalesce into one solid frozen mass. This mass may be a ring or donut as needed to support a shaft or a solid block of whatever shape necessary to stabilise the working face or heading.

Because of the dearth of engineering data on the properties of frozen ground (especially clays) it is recommended that two steps be taken early in any design of ground freezing:

1. A qualified consultant be engaged to advise on the design and construction of the project. Advice from such a professional is essential for the work and will pay for itself many times over.
2. Laboratory tests be designed and carried out using soil samples from the actual site. Only in this manner can meaningful properties of frozen soil be obtained for the site involved for purposes of conceptual engineering ("scoping the problem").

However, a few general guidelines can be stated as follows (after Xanthakos, 1994).

1. Pipes are normally spaced 3 to 4 feet apart.
2. Select a spacing-to-diametre ratio <13 (for pipes 120 mm or less in diametre).
3. Use a brine temperature <25ω C.

4. Provide 0.013 to 0.025 tons of refrigeration per foot of freeze pipe.
5. Determine typical frozen ground properties by laboratory testing.

Groundwater flow across the site requires special considerations closer pipe spacing, multiple rows of pipes and the like. Groundwater flow velocities approximately >2 m/day may impede or prevent freezing. A number of special challenges associated with ground freezing should be considered in both the design and construction stage. Those are creep of frozen ground, sensitivity of frozen ground properties to loading condition, ground heave or settlement, and others.

Tunnelling in Difficult Ground

Engineers like to work with materials having defined characteristics that do not change from one location or application to another. Unfortunately, geology seldom if ever cooperates with this natural desire but instead tends to present new and challenging conditions throughout the length of a tunnel. Some of these conditions approach the "ideal" closely enough that they can be approached as presented for rock and soft ground. However, in many cases special approaches or arrangements must be made to safely and efficiently drive and stabilise the tunnel as it passes through this "Difficult Ground".

The factors that make tunnelling difficult are generally related to instability, which inhibits timely placement or maintenance of adequate support at or behind the working face; heavy loading from the ground which creates problems of design as well as installation and maintenance of a suitable support system; natural and man-made obstacles or constraints; and physical conditions which make the work place untenable unless they can be modified.

Instability

Instability can arise from: lack of stand-up time, as in non-cohesive sands and gravels and weak cohesive soils with high water content or in blocky and seamy rock; adverse orientation of joint and fracture planes; or the effects of water.

The major problems with mixed face tunnelling can also be ascribed to the potential for instability and this class of tunnelling will be discussed under this heading.

Heavy Loading

When a tunnel is driven at depth in relatively weak rock, a range of effects may be encountered, from squeezing through popping to explosive failure of the rock mass. Heavy loading may also result from the effects of tunnelling in swelling clays or chemically active materials such as anhydrite. Adverse orientation of weak zones such as joints and shears can also result in heavy loading, but this is usually dealt with as a problem of instability rather than loading. Combinations of parallel and intersecting tunnels are a special case in which loadings have to be evaluated carefully.

Obstacles and Constraints

Natural obstacles such as boulder beds in association with running silt and caverns in limestone are just two examples of natural obstacles that demand special consideration when tunnelling is contemplated. In urban areas, abandoned foundations and piles present manmade obstructions to straightforward tunnelling while support systems for existing buildings and for future developments present constraints which may limit the tunnel builder's options. In urban settings, interference conflicts, public convenience or the constraints imposed by the need or desire for connection to existing facilities will sometimes result in the need to construct shallow tunnels, which have a range of problems from working in confined spaces, avoiding subsidence and uneven ground loading and support.

Physical Conditions

In areas affected by relatively recent tectonic activity or by ongoing geothermal activity, both high temperatures and noxious, explosive or deadly gases may be encountered. Noxious gases are also commonly present in rock of organic origin; and elevated temperatures are commonly associated with tunnelling at depth. In an urban setting, contaminated ground may be encountered and will be especially troublesome when found in association with other difficult conditions.

Where appropriate, some information is provided as to the reasons why the condition under discussion creates problems for construction. Some examples of each of the conditions referred to above are discussed briefly to yield insight into the problems and to define the range of solutions available.

Instability

Non-Cohesive Sand and Gravel

Cohesion in sands is more than a matter of grain size distribution. For instance, beach-derived sands normally contain salt (unless it has been leached out), which aids in making sand somewhat cohesive regardless of grain size. The moisture content then becomes a determining factor.

The age and geologic history of the deposit is also important since compacted dune sands with "frosted" grain surfaces may develop a purely mechanical bond; and leaching and redeposit of minerals from overlying strata may also provide weak to strong chemical bonding.

If groundwater is actually flowing through the working face, any amount may be sufficient to permit the start of a run which can develop into total collapse. There is no such thing as a predictably safe rate of flow in clean sands. Uncontrolled water flows affect more than the face of the excavation. If the initial support system of the tunnel is pervious, water flowing behind the working face will carry fines into the tunnel and may create substantial cavities—sometimes large enough to imperil the integrity of the structural supports. This phenomenon occurred in Los Angeles where a ruptured water main caused sufficient flow through a tunnel support system to cause a failure and resulting large sink hole in the street.

While factors such as compaction or chemical bonding may permit some flow without immediate loss of stability, this is not a reliable predictor. Soil deposits are hardly ever of a truly uniform nature. It has been observed in soft ground tunnels in recent deposits that all that is necessary to trigger collapse may be the presence of sufficient water to result in a film on the working face; i.e., there is no negative pore pressure to assist in stabilising the working face. Of course, there is never a safety factor arising from surface tension (capillary action) in coarse sand or gravel.

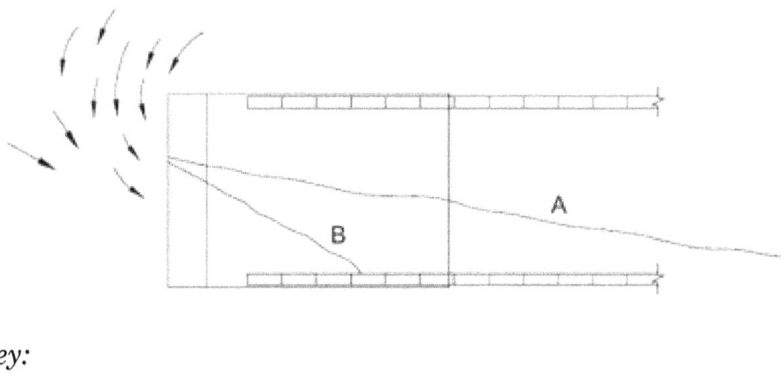

Key:

A: Flowing Sand

B: Sand at Normal Angle of Repose

Figure: Flowing Sand in Tunnel

The cleaner the sand, the more liable it is to run or flow when exposed in an unsupported vertical face during tunnel construction. Single sized fine grained sands (UCS classification SP) are the most troublesome, closely followed by SP-SM sands containing less than about 7% of silt and clay binder. Saturated sands in these classes have been observed to flow freely through sheet piles and to settle into fans having an angle of repose of less than 5. Unconfined SP sands will run freely, as in an hourglass, whether wet or dry, having some stability only when damp but less than saturated (no piezometric head). The large proportion of the sand particles of the same size allow the sand to move almost as freely over one another as would glass marbles.

Silt, intermediate in grain size between sand and clay, may behave as either a cohesive or non-cohesive material. In some areas it is common to find thin seams of saturated fine sandy silt trapped between clay beds in glacial deposits. In general, unless the seams are thicker than about 9-12 inches, when the silt layer is exposed in the wall of an excavation, the soil slumps out at intervals leaving a series of small shallow caves like entrances to burrows. The water appears to drain fast enough from the increased surface area exposed so that the remainder of the exposed material stabilises.

The usual problem encountered with running sand is settlement and cratering at the surface with damage to structures or utilities in the area. If the ground is permeable, consolidation grouting of the entire sensitive area can be undertaken to stabilise the soil before tunnelling. If dewatering is successful in depressing the water table below the tunnel invert, it may be found that the sand is just as unstable dry as wet. The alternative of using compressed air is attractive, provided the working pressure is very carefully controlled; but even so, the ground may be too dried out for stability.

If the face is a full face of sand and similarly weak materials, a slurry machine or an earth pressure balance machine, will be required. In general, rotary head tunnelling machines for soft ground tunnels require very similar physical properties over the entire working face and the entire job. If these conditions do not prevail, then weaker ground, and running sands in particular, must be prevented from entering the shield more rapidly than is proper for the rate of advance. Slurry shields have the best opportunity of controlling variable conditions where running sands are present; but they will prove difficult to keep on line and grade in mixed face conditions if one of the beds present is even a strong clay. If the sand and clay beds are more or less evenly distributed (e.g., a varved clay), then this problem may not arise. Of the digger type shields, neither extensible poling plates nor orange peel breasting have proved to be generally successful, hence these machines are now rarely used.

A problem with all shield construction is the necessary difference in diametre between the shield and the lining. If the soil has no stand-up capability by the time it is exposed in the upper part of the tunnel before expansion of a primary lining or introduction of pea gravel or more commonly, grout into the annular space for non-expanded linings, then there will be loss of ground. If the unfilled annular space averages one inch in a 20 ft tunnel, the lost ground from this single cause is approximately 1.7% shown in Table 7.2 as "poor" practice. Even if only local ravelling takes place, it may choke off the flow of grout before the void can be filled with a continuous supporting fill material. This loss of ground results in a contribution to settlement.

Soft Clay

For the purposes of this discussion, soft clay includes any plastic material that will close around a tunnel excavation if free to do so. This will be the case if the overburden pressure at spring line exceeds the shear strength of the clay by a factor of about three or more. However, if the clay is sensitive and loses strength when remolded, the remolded strength will govern some of the clay behaviour during tunnel construction. The phenomenon of sensitivity is mediated by several factors that cannot be fully discussed here but, in general, sensitivity may be suspected in clays with a high moisture content. Particularly at risk are marine clays from which the salt has been leached. The loss of strength may lie within a wide range, the ratio of undisturbed to remolded strength sensitivity being from 2 to 1,000. Moderate sensitivity of 2 to 4 is quite common. During remolding, the void ratio in the clay is reduced and free water is released. When this free water has access to a drainage path such as a sand bed or the tunnel itself, there will be a volume change in the soil mass which will result in surface settlement.

The cohesive soil is to be stabilised so that closure around the tunnel lining is minimised and stable control of line and grade are maintained, the critical number must be reduced below about 5; this will enable reasonable control of alignment and grade. The following equation:

$$P_a = P_z - (N_{crit} \; x \; S_u)$$

where Ncrit is the critical number, Pz is the overburden pressure at tunnel spring line, Pa is the working pressure in a compressed air tunnel or the equivalent average pressure provided by the initial support system, and Su is the undrained shear strength of the soil in compatible units. As an example, if N is to be maintained at a value of 5, the overburden pressure is 40 psi and the unconfined shear strength of the soil is 1,000 psf = 7 psi, the required working pressure in the tunnel will be (40 - 5 7) = 5 psi. It can be seen that if the shear strength of the soil is reduced by remolding caused by passage of the shield through the ground to a value of 250 psf, then the required air pressure for stability increases to over 30 psi, transforming the project from a relatively straightforward one to a difficult one.

Attempting to calculate the required volume of grout injection into the annular void between shield excavation and lining in clays often is not a fruitful exercise. It will certainly be possible to inject the requisite volume of grout, but it may be difficult to make it flow around the tunnel perimetre in an even layer. The best results are obtained by establishing multiple simultaneous injection points permanently fixed within the shield tail and passing through the tail seals. Grout is injected throughout the time the shield is in motion. For this system to work, the lining must be a bolted segmented lining with built-in gashets between segments. It must be expected that for simultaneous injection through multiple ports while the shield is in motion there will be a substantial learning curve before all elements of the system are functioning smoothly to achieve the desired result.

It is generally difficult to use any mechanical excavation equipment in this type of ground except for a slurry shield or earth pressure balance shield (EPB). These days, the two types of machine are approaching interchangeability with the continuing development of chemical additives (conditioners). The edge goes to slurry machines in coarse geology and/or where the rock crusher may be needed to reduce rock or boulders to a size that will pass the machine.

The EPB is preferred as being somewhat more flexible in varying conditions and somewhat less expensive than a slurry shield. In order to control pressure in the plenum chamber behind the cutterhead, a screw conveyor is required. The rotational speed of the screw is matched to the advance rate of the EPB and pressure in the plenum is monitored using multiple sensors. If boulders are likely to be encountered, especially if they will be larger than can pass through the screw conveyor, the cutterhead must be fitted with disk cutters in addition to the drag bits normally associated with this type of machine.

Blocky Rock

Rock is a basically strong material which requires little or no structural support when intact; although it may require protection from exposure to air, water or from fluids conveyed in the tunnel. However, when the rock joints and fractures are open sufficiently that the natural rugosity of the block surfaces will not prevent movement of rock blocks or substantial fragments, the rock is said to be "blocky." If the joints and fractures contain clay-like material resulting from weathering or light shearing, then the rock is described as "blocky and seamy." As can be seen from Table 6-7, this may raise the rock load by a factor of approximately three. In zones where the rock has small folds, but is open along the direction of the folds, it may be free to move in only one direction. Such rock is still blocky.

When rock is subjected to the action of explosives, high-pressure gases flow into any fissures in the rock before they have finished their explosive and rock-fracturing expansion. Even in hard granite, a result of blasting is the creation of micro-fissures extending well outside the blasted perimetre. In blocky rock, the effect may well extend more than a tunnel diametre outside the desired finished surface; a good deal of overbreak and potential loosening and movement of blocks is likely to result.

Another problem with this type of rock is that it is highly susceptible to the destabilising effects of water flowing through the fracture system with sufficient energy to dislodge successively more rock. Finally, it is quite likely when blocky and seamy rock is encountered in a tunnel excavation, especially in heavily folded strata, that there will be zones where the weathering has proceeded to a conclusion resulting in the presence of weak earth-like material with little capacity to sustain loads or to preserve the tunnel outline.

All of the rock conditions described require early and carefully placed primary support to preserve ground stability and to provide a safe workplace. Even before support installation, it is necessary to minimise surprises by scaling off any loose rock which will present a hazard to the crews installing the support system. Many still prefer to use steel ribs and wood lagging in this type of rock. It provides positive support and is quickly installed in tunnels less than about 5 metres in diametre. Unfortunately, crews still have to work under the unsupported rock to install the ribs and lagging; the material costs are high; the presence of timber results in the possibility of future uneven loading on the permanent tunnel lining as wood rots out and steel corrodes; and it becomes relatively difficult to ensure good contact between the lining concrete and the rock even after contact grouting.

For these reasons the use of shotcrete and rock bolts has become popular. In rock known to be blocky and therefore to need support, an initial layer of shotcrete about 5 cm thick should be applied as soon as possible in the tunnel crown. This is followed by the installation of pattern rock bolts whose length and diametre are governed principally by the tunnel diametre.

Adverse Combinations of Joints and Shears

Jointing systems in rock arise from many causes, some of which are noted here. Sedimentary rocks, and particularly limestone, typically have three more or less orthogonal joint sets arising from the modes of deposition and induration which formed them. Not all joints are continuous, but those in any set are parallel. There may be many sets or, in weak, massive sandstone, for instance, only one or two. Joints and fracture systems combine to break up the rock mass into interlocking fragments of varying sizes and degrees of stability.

In the absence of direct evidence to the contrary, it should be assumed that shears and faults are continuous throughout their intersection with the tunnel excavation. In schistose materials, weathering usually follows a foliation plane to great depths, even in temperate climates when a weak zone has been formed by slippage along that plane. Other faulting may cause the development of extensive fracture systems in any direction. A section through the project area perpendicular to the strike of the exposed surfaces in schistose materials will generally reveal a saw-tooth profile with one of the surfaces parallel to the foliation. Continuation of the plane thus defined to tunnel elevation will be a preliminary indicator of the presence of sheared and weathered rock in the excavation.

Continuous joints and shears can define large blocks with little or nothing to hold them in place once the tunnel excavation has been completed. It is important to identify the locations of blocks with the potential for falling out in order to provide support during cautious excavation. For large diametre tunnels in particular, this requires an assessment of the potential before construction begins, mapping during construction, and control of drift size and round length to ensure against complete exposure of an unstable block in a single round.

The difficulty of controlling the correct placement of steel sets in multiple drift headings works against the use of this kind of support. Initial rock bolting followed by reinforced shotcrete is a reasonable approach. In all cases where rock bolts have to be located to take direct and reasonably predictable loads, it is better that they be installed ahead of the shotcrete while the joint locations are still visible. If mechanical rock bolt installers cannot be used, then the crews must be protected by overhead cages.

Faults and Alteration Zones

Tectonic action, high pressure and high temperatures may metamorphose rock into different structures with unpredictable joint patterns. The uplift and folding of rocks by tectonic action will cause fracturing perpendicular to the fold axis along with faulting where the rock cannot accommodate the displacements involved, so that shears develop parallel to the fold axis. Other types of faults arise as the earth accommodates itself to shifting tectonic forces. Faults or shears may be thin with no more significance than a continuous joint or they may form shear zones over a kilometre wide in which the rock is completely pulverized but with inclusions of native rock, sometimes of large size.

All of the conditions briefly described above may be additionally complicated by the presence of locked-in stress, high overburden loads, or water.

Dealing with the conditions encountered in such fault zones and weathered intrusive zones depends on the excavation method in use, the depth below the ground surface, the strength of the fault gouge, the sheared material or the weathered or altered rock, and the water conditions. Water problems are discussed in general in the next section, including consideration of the difficult water conditions commonly found in association with faults; however, to the extent that they affect the selection of construction methods appropriate to fault crossings, they are referred to here.

Current technology provides other solutions, such as the use of precast concrete lining in the weak ground with supplementary jacking capability to enable the lining to provide the jacking reaction for the thrust of the TBM.

In general, fault crossings offer conditions akin to those of mixed face tunnelling and the same methods are available to deal with them. Different circumstances come into play with deeper tunnels, especially if these are of large diametre. Such tunnels are usually long and logistics are important. The comparative lengths of fault zone and normal tunnel dictate that the construction method be efficient for the normal tunnel. Nevertheless, sufficient flexibility is required to permit safe and reasonably expeditious construction through the worst conditions likely to be encountered. Drill and blast excavation is still commonly used in such tunnels. Rock bolts and shotcrete then become the preferred support system, although steel ribs and lagging or steel ribs with shotcrete are also still used. TBM successes in these conditions have been few. There are two principal problems: the loose material in the fault runs into the buckets and around the cutters and stalls the cutterhead; and if the fault contains cohesive material, it squeezes and binds the cutterhead and shield with similar results.

One solution to the problem of loose or loosened ravelling and running material is to establish a grout curtain ahead of the TBM and then to maintain it by continuing a grout and excavation cycle throughout the fault-affected portion of the drive. Even if imperfect—as consolidation grouting tends to be, especially when placed from within the tunnel in conditions providing limited access—it is likely that a properly designed and executed program will add sufficient stability to the ground to permit progress. It should be noted that any such program will be expensive and time-consuming. It is therefore unlikely that any contractor will willingly do the necessary work unless it has already been envisaged in the contract as a priced bid item. It is also important to recognise that if water is running into the tunnel through the working face, a bulkhead will be required to stop the flow while the initial grouting is in progress. Grouting into running water is a slow and expensive way to establish a grout seal.

Within limits, the squeezing problem can be dealt with in part in TBM tunnelling by tapering the shield and making its diametre adjustable within limits; and by bevelling the cutterhead itself to the extent that this is possible without interfering with the efficiency of the buckets. Expandable gauge cutters are also used, but this is still a developing technology. One of the problems is that there is a tendency for local shearing of the cutter supports to result in an inability to withdraw the cutter once it has been extended. Also, since such cutters are acting well outside the radius of the buckets, muck which falls to the invert is not collected but provides an obstruction the cutters must pass through repeatedly. This grinds the debris finer and finer and abrades the cutter mounts as well as the cutter disk. This makes it necessary to provide means for eccentric cutterhead rotation so that the invert is properly swept. Unfortunately, squeezing is commonly, if not most often, manifested preferentially in the tunnel invert.

Water

It was Terzaghi's view that the worst problems of tunnelling could be traced to the presence of water. Among other things, he considered that (except for circular tunnels) it was prudent to double the design rock load on the tunnel lining when the tunnel was below the water table. This in itself would not be a serious problem, since most tunnel linings are already limited as to their minimum dimensions by problems of placement rather than by design considerations. However, there are many other problems that are associated with the presence of water. Several are discussed below, working in sequence from clay to rock and, within rock, from weak and fractured to strong and intact.

Clay

Most clays are at least slightly sensitive. This arises from the microstructure of clay soils which are composed largely of platy minerals. As with a heap of coins, the packing is not perfect, even though the clay is relatively impermeable. Each fragment is held in place by some combination of free body equilibrium forces, ionic interaction and chemical or mechanical bonds at the contact points. The pores of the clay are generally filled with water, which may contain salts in solution. Disturbance of the clay results in disruption of the bonding, migration of water and at least temporary weakening of the clay structure. The free water will be released at any temporary boundaries formed by shearing. As the clay reconsolidates, it is likely to gain strength over the initial condition, but this will be a protracted process.

The immediate effect, and the one that affects tunnel construction, is loss of shear strength throughout the disturbed mass. In organic silty clays, the sensitivity is commonly about 4, indicating a fourfold loss of strength upon remolding. This is associated with an initial water content of about 60%. As shown on Page 8-4, a four fold loss of strength can result in more than a sixfold increase in the required support. In any one material, the sensitivity may vary greatly, depending on the water content. Sensitivities as high as 500 to 1,000 may be found in some clays, such as the Leda clay commonly encountered in previously glaciated areas. Marine clays such those found in Boston lose salt by diffusion when situated below the water table. Such clays are typically highly sensitive.

Tunnelling is already sufficiently challenging in moderately sensitive clays as the critical number suffers a local fourfold or more increase. For shielded tunnelling, it is very important to avoid excessive efforts to correct line and grade as it is easily possible to create a situation in which control is lost.

A further effect of disturbance of sensitive clays is directly dependent on the loss of pore water expressed from the clay. The volume change results directly in rapid subterranean and surface settlement. In addition, the clay closes rapidly on to the tunnel lining, resulting in even greater settlement unless sufficient compensation and/or contact grout can be injected promptly.

Mixed Face Tunnelling

Tunnelling in mixed face conditions is a perennial problem and fraught with the possibility of serious ground loss and consequent damage to utilities and structures as well as the prospect of hazard to traffic. The term "mixed face" usually refers to a situation in which the lower part of the working face is in rock while the upper part is in soil. The reverse is possible, as in basalt flows overlying alluvium encountered in construction of the Melbourne subway system. Also found are hard rock ledges in a generally soft matrix bed of hard rock alternating with soft, decomposed and weathered rock; and non-cohesive granular soil above hard clay and hard, nodular inclusions distributed in soft rock (e.g., flints beds in chalk or garnet in schist).

The primary problem situation is the presence of a weak stratum above a hard one as clearly illustrated for the construction of the 2.3 km long C line and the 4 km long S line of the Oporto Metro project as a part of the mass transit public transport system of Porto, Portugal (Babendererde et al., 2004). The highly variable nature of the deeply weathered Oporto granite overlying the sound granite posed significant challenges to two 8.7 m diametre EPB Tunnel Boring Machines.

There will always be water at the interface which will flow into the tunnel once the mixed face condition is exposed. This increases the hazard because of the destabilisation of material already having a short stand-up time. Stabilisation therefore calls for groundwater control as well as adequate and continuous support of the weak material. Moreover, this support must be provided where energetic methods, such as drill-and-blast excavation, are required to remove the harder material.

Dewatering can reduce the head of water, but it cannot remove the groundwater completely; nor can it be realistically expected to offer control on an undulating interface with pockets and channels lower than the general elevations established by borehole exploration. Compressed air working will not deal with water in confined lenticular pockets and it is usually inappropriate when the length of the mixed face and soft ground conditions amount to only a few percent of what is otherwise a rock tunnel. Also, recent experience where extensive beds of clean (SP and SP-SM) sands have been major components of the weak ground shows that compressed air alone will not stabilise the ground which becomes free-flowing as soon as it has dried out. Therefore, on the whole, consolidation grouting is to be preferred in this situation.

It is emphasized that the best time to seal off groundwater is before it has started to flow into the tunnel. Once the water is flowing, it is extremely difficult to stop it from within the tunnel except by establishing a bulkhead.

Tunnelling in Difficult Ground

Heaving Loading

Squeezing Rock: When a tunnel opening is formed, the local stress regime is changed. The radial stress falls to zero and the tangential stresses increase to three times the in situ overburden load (neglecting the effects of any locked-in stress resulting from past tectonic action that has not been relieved). If the unconfined compressive strength of the rock is less than the increased tangential stress, a mode of failure will be initiated which is described as "squeezing rock". As elastic failure occurs, with consequent reduced load-bearing capacity of the ground, the load is transferred by internal shear to adjacent ground until an equilibrium condition is reached. If the ground develops brittle failure and is shed from the tunnel walls, then there will be no residual strength of the failed ground to share in the load redistribution. If the ground is sufficiently weak or the overburden load too great, the unrestrained tunnel may close completely.

The Squeezing Process

The detailed mechanism of ground movement is complex and depends on the presence or absence of water and swelling minerals as well as on the physical properties of the ground. For the purposes of this discussion, however, the squeezing process may be described as follows.

Initial Elastic Movement

As the tunnel is excavated, stress relief allows elastic rebound of ground previously in compression to relieve stress. This stress relief occurs beyond the working face as well as around the tunnel excavation. In thinly laminated rocks such as schist and phyllite, the modulus of elasticity parallel to the foliation is likely to be much higher than that in the perpendicular direction. Therefore, the elastic movement immediately distorts the shape of the excavation as the rock moves a greater distance perpendicular to the foliation than parallel to it. Moreover, since the rock can move more easily along regular foliation planes than perpendicular to them, more than one factor is at work determining the actual distortion of the tunnel shape. The elastic rebound takes place in all tunnel excavations and is not properly a part of squeezing, which is associated with changes in the rock structure. However, the associated increase in tangential stress in the rock initiates the next phase of movement (squeezing) as the rock fails. As the rock moves toward the tunnel opening, the circumference of the tunnel shortens. There is a limit imposed by the modulus and strength of the rock on how far this process can continue before elastic failure is initiated. Consider a rock of compressive strength 35 Mpa and an elastic modulus of 17,500 Mpa. The circumferential strain per unit length at failure will be 35/17,500 cm/cm or 2 mm/m. For a tunnel of 2 m radius therefore, a shortening of this radius by about 4 mm implies the initiation of impending elastic failure at the exposed rock surface. This does not mean that the rock suddenly loses all strength (unless it is brittle enough to flake off the wall) but rather that its residual strength is greatly reduced. As the tangential shear stress builds up there will come a time when the differential stress is sufficient to cause internal shear failure. This is manifested by the development of new parting surfaces where the overstressed rock separates from the neighbouring rock.

Strength Reduction

When the rock remaining is insufficiently strong to carry the increased load passed to it as shearing progresses it will fail in turn. In strong and brittle rocks, this failure can result in explosive release of rock fragments from the surface in a phenomenon known as "rock bursting." A somewhat gentler expression of the same phenomenon is known as "popping rock," which is still a dangerous phenomenon. Because these occurrences actually remove rock from the surface there is obviously no residual load-carrying capability of the failed rock. In weaker and less brittle rock the failed material stays in place and enters a plastic or elasto-plastic regime. Its modulus of elasticity and its unconfined compressive strength (which represent its load-carrying capacity) may be reduced by two orders of magnitude, but it can still support some load. In the meantime, the load shed by the failed rock at the perimetre of the opening is transferred deeper into the rock mass where the degree of confinement is higher and the ultimate load-bearing capacity is therefore also higher. The phenomenon may be modelled step-wise, but it is truly a continuous process and will cease only when the total load has been redistributed. Depending on the amount of excess load-carrying capacity available in the partially confined rock around the tunnel perimetre, the stress regime may be affected up to several tunnel diametres away from the opening.

Compounding the stress increase, which leads to failure, is the similar regime in the dome ahead of the working face. The abutment of this dome is the already overstressed rock behind the working face. The problem is therefore three-dimensional in the region affected. The initial movements associated with strength reduction take place quite fast, so that as much as 30% of the final loss of tunnel size may be completed within one to one and a half tunnel diametres behind the working face.

Creep

As a consequence of the reduced elastic modulus and the reduced strength of the rock additional radial movement of the tunnel walls occurs. In the zone outside the tunnel, the rock properties are substantially changed. In particular, both the elastic modulus and the unconfined compressive strength decrease continuously (but not in a linear fashion) from their original values still existing in undisturbed rock toward the tunnel wall. The tunnel decreases in diametre as the weakened material creeps toward the tunnel boundary. The rate of movement is roughly proportional to the applied load. The movement is therefore time-dependent (after the initial elastic stress relief, which may be regarded as essentially instantaneous). As the ground is allowed to strain, so the strength of the support required to restrain further movement is reduced. However, depending on the amount of squeezing, shear failures and dilatation accompanying failure may result in unstable conditions in the tunnel walls and crown. Since the timing, location and amount of such failures are not subject to precise definition, support is usually introduced well before the full amount of potential movement has occurred.

Modelling Rock Behaviour

Because of the nature of the failure mode, elasto-plastic and visco-elasto-plastic mathematical models have been developed to describe the resulting movements and to evaluate the stress regimes for tunnels in rock. These models are not exact but correspond sufficiently well with experience to be useful. Unfortunately, for any given tunnel they depend on the use of information which can only be derived from experience in the specific tunnel involved.

It has been noted from experimental work that the net load appearing at the tunnel surface varies with the tunnel diametre as a power function. The loading is also dependent on the rate of tunnel advance. It is therefore clear that when such conditions are encountered, the smallest tunnel diametre adequate for the purpose should be selected. Experience also shows that circular tunnels are easier to support than any other shape.

Other Factors

If the rock contains porewater, negative pore pressures are set up as the rock moves toward the tunnel. This provides limited initial support until the negative pore pressure is dissipated. In addition, the new pressure gradient set up by the release of confining pressure results in seepage pressures toward the tunnel boundary. In regions of high hydrostatic head, significant increases in rock loading can occur. It is also thought that even small proportions of swelling clay minerals in the rock can contribute significantly to rock loads when water is present. This water need not be flowing—only present in the pores. When all factors contributing to rock mass behaviour have been identified and quantified, it may be possible to develop more exact predictive models and to devise new means for controlling and improving ground behaviour. In the meantime, we must make do with approximations based on experience.

Monitoring

Rate of squeeze and rock loads are somewhat dependent on tunnel size and rate of advance. It is essential in squeezing (or swelling) conditions—or even in blocky and seamy rock where joint closure may create problems—to establish a program of convergence point installations which will be routinely used to monitor the amount and rate of movement of the tunnel walls. This information collected over time and collated with the behaviour of the tunnel support system will provide the information needed both to predict and to install the appropriate amount of support as tunnelling progresses. This technique lies at the heart of SEM tunnelling in rock.

Yielding Supports

One approach to squeezing rock is to go to a simple and workable system of yielding supports as illustrated. The number of yielding joints can be modified to provide the needs of the rock currently being excavated since all components are manufactured on site. Each joint permits up to 22 cm of closure. It has been found essential to shotcrete the gaps once the closure nears the limit allowed without the steel sections actually butting together. Failures have been common when this butting has been allowed to happen. It has also been found that allowing the invert to heave freely for twenty to thirty days before making an invert closure allows the total support system to resist all remaining loads with some reserve capacity for long term load increases. Other, more complicated yielding systems have been designed and used.

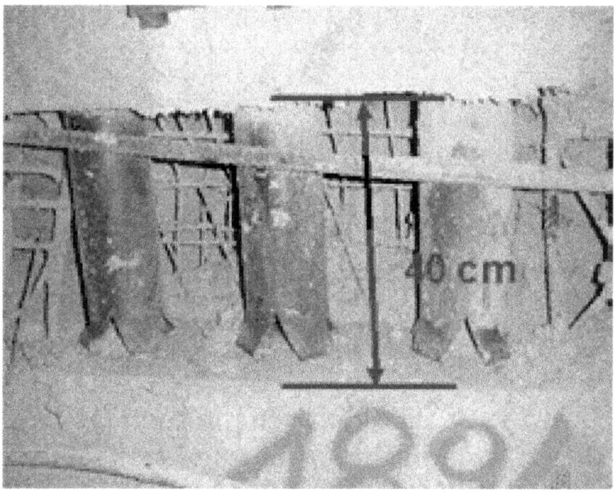

Figure: Yielding Support in Squeezing Ground

The support system provides a relatively low initial support pressure and permits almost uniform stress relief for the rock in a controlled manner around the entire circumference of the tunnel while preventing the rock from ravelling. The shotcrete is not damaged by the convergence because of the yielding joints and so maintains its integrity, provided that timely closures are made. After allowing practically all of the stress relief required by the elasto-plastic stage of rock deformation, the support system is made rigid whence it can support a pressure of 3.8 MPa which is available to deal with long term creep pressure.

Timber Wedges and Blocking

The use of blocking tightened against the ground by pairs of folding wedges introduces a structural element that can be allowed to fail by crushing. Observation of progressive failure coupled with experience provides warning that the behaviour of the steel supports should be closely monitored in case a decrease in spacing or an increase in section becomes necessary.

Precast Invert

When squeezing is sufficiently severe to be troublesome. It will be seen that the prime tendency is for the lower sidewalls to move in and for the invert to heave. The loss of strength of the rock in the invert leads to the rapid development of muddy and unstable conditions under the tunnel haulage operations. In this case it may be desirable to use precast concrete invert slabs kept up close to the working face in place of invert struts. In one location where high squeezing occurred, such slabs were heaved up and required maintenance to keep the track on grade. However, they did provide a good and trouble-free surface otherwise.

TBM Tunnelling

Because of the number of large tunnels now under consideration where the use of TBMs is contemplated and where squeezing conditions may become important, the following discussion is extended, even though not based on a great deal of current experience.

The majority of examples of tunnels in squeezing ground are related to the crossing of faults. TBMs have been troubled in this situation by inrushes of water carrying sand and finely divided rock or by blocks of rock jamming between the cutters. The second of these problems has been dealt with in many tunnels in otherwise normal conditions. The primary solution is the use of a machine design which allows only a limited projection of the cutters forward of the cutterhead by means of a face shield ahead of the structural support element. The second development is a design which permits worn cutters to be changed from within the tunnel, so that no access is required in front of the cutterhead. There has, as yet, been no easy solution for the problem of the cutterhead and its buckets being choked with sand and rock fragments while unrelenting water flows are in progress. It becomes a difficult and slow process of cleaning out and gaining progress slowly until the affected area has been cleared. Also in such conditions, the presence of a shield is important to protect the machine and to provide temporary support to material with no stand-up time. In some circumstances, if the condition is known to exist or to be likely to exist, probing ahead to identify the precise location can give an opportunity to stabilise the ground with grout injections, keeping a bulkhead thickness ahead of the excavation at all times.

It is sometimes possible to allow most of the water to drain out of the ground, but this is not a reliable approach to prediction of construction methods. Shielded TBMs have been used successfully in such conditions, but unfortunately the use of a long shield militates against successful use in squeezing ground.

The other major problem, whether or not in a fault or shear zone, is the closure of the ground around the cutterhead shield and any protective shield behind the cutterhead. Many TBMs, have been immobilised because the load on the shield system was too high to permit the machine to advance. One way to approach this problem is by the use of a short shrinkable shield on the machine.

It is not anticipated that tunnelling in squeezing ground or fault zones will ever become a simple and routine operation because of the erratic and unpredictable variability of conditions. However, the current climate of opinion is that virtually all tunnels can be attacked by TBM methods whenever there is an economic advantage in doing so.

As previously noted, the difficulty of predicting rock behaviour in squeezing ground has played a major role in the development of observational methods for determination of rock support requirements. However, if tunnelling by TBM is selected, some of the flexibility of the observational method is removed because it is difficult to see the face or to measure movements Hence, decisions must be made at the time the TBM is designed as to the amount of ground movement to be anticipated or permitted and the design of the support system to accept the loadings implied at different stages of the tunnelling operation.

Since squeezing of soft rock does not usually lead to immediate instability, it should be possible and practical to delay major support installation until a high percentage of the total strain has taken place and ground loading has been reduced. Sixty to seventy percent of the potential ground movement has usually taken place within about three diametres of the working face. If the total amount of squeezing is not great, it may not be necessary, or even desirable, to delay support installation so long.

Ideally, final support is not installed until convergence is less than one millimetre per month. The loading associated with a given amount of convergence is dependent on the parameters of the project. It is also important to take into consideration any long-term requirement for the tunnel to carry water. Finally, it must be realised that if groundwater is to be totally excluded from the tunnel, the final lining must be designed to carry the full hydrostatic head unless the aquifer is fully sealed off by consolidation grouting. If groundwater is admitted, whether in a controlled manner or by allowing local cracking of the lining, then only seepage pressures need be accounted for. In the case of weak squeezing ground or faulted rock with an unknown potential for swelling behaviour, the latter alternative appears undesirable.

Steel Rib Support System

Steel ribs set close to the tunnel surface and blocked from it are often used as the initial support system for rock tunnels especially those constructed by conventional drill-and-blast methods. Wood, concrete or steel lagging may be placed between the ribs and the rock to secure blocky or ravelling ground or welded wire fabric may also be used. The same system can also be used in TBM tunnels, but it is necessary to allow an initial small distance between rib and ground so that the last rib segment can be positioned conveniently. In normal tunnelling, this space is later closed by expanding the rib against the ground. Especially in squeezing ground, the rib must be blocked to the rock all around its perimetre. As the ground movement occurs and continues, it will squeeze past the ribs and stress relief will occur. In this type of installation, it is necessary for the ribs to be as stiff as possible to prevent displacement and buckling. The chief safeguard is to install steel ties and collar braces at intervals around the rib. The collar braces are typically steel pipe sections set between the ribs. The ties then pass through holes in the web of the steel section and through the pipe forming the collar brace. These members are also subject to deformation by the invading ground. If this creates any substantial problem, angle irons welded to the inner face of the ribs can be substituted.

It is significant that in tunnels where the ribs have buckled under squeezing load but have been left in place, they commonly retain enough structural strength to provide support. The problem is that the squeezing usually intrudes on the required final profile of the tunnel.

Concrete Segments

Segmental concrete linings take two quite different forms. The traditional bolted and gasketted lining is meant to be a final lining erected in one pass. Until recently, the more common application has been to use unbolted, ungasketted segments, with light reinforcement to allow handling, as a "sacrificial: primary lining. This latter type of lining is sacrificial only in the sense that it is allowed to sustain fractures resulting from jacking loads or redistribution of stress; it retains most of its initial load-bearing capacity. A final lining is always placed within this type of lining; it sometimes is an unreinforced concrete lining of nominal thickness, say 10 inches. The combination lining may be less expensive than the one-pass system and has the merit of flexibility. Problems arose with the precast concrete tunnel lining when there was insufficient erection space to allow for deviations normal to tunnelling.

In electing to use a precast concrete lining decisions are necessary as to the amount of ground movement to be allowed and the backfill material to be used between the lining and the rock. In allowing for a large amount of potential ground movement, certain problems of erection stability arise. The lining will require support clear of the invert and a horizontal tie or blocking to keep it in shape during and after erection until backfill grouting is complete. There is time and skill involved in executing the work, but no significant difficulty.

Current technology is now trending towards the use of a one pass system of concrete segments. These segments are of high quality concrete and are usually bolted and gasketed at all joints. However, specially doweled circumferential joints are being used. It is necessary that such rings be cast and cured in a controlled factory environment and that they be of high strength concrete for high resistance and high elastic modulus. Steel fibers may be used in lieu of reinforcing steel in some applications.

It is important that the moving ground should not come into contact with the completed ring at a point. Distortion would necessarily result with a possible consequence of reducing load-bearing capacity. It is also possible to use compressible backfill in the annular void provided the material offers sufficient resistance to mobilise passive reactions sufficient to withstand distortion of the lining. At the least, careful consideration would be needed in specifying the strength and deformability of any compressible material to be used.

TBM Tunnelling System

Cutterhead

Many different cutterhead designs have been used over the years from the earliest flat heads with multiple disc cutters through domed heads, rounded edge flat heads and conical designs. These days the cutterhead geometry is selected on the basis of the ground it is expected to penetrate. It has been found preferable to arrange that at least the gauge cutters be designed to be changed from behind and it is now common to arrange this system for all cutters. A spoke design allows ready access to the working face and simplifies design in some respects. However, such machines offer little support if weak ground is encountered and it is generally considered prudent to use a closed face machine. Also, to protect the cutters and cutter mounts, a lighter false face is provided so that the cutter disks protrude only a short distance.

In conventional designs, the cutterhead is provided with its own shield as part of the cutterhead bucket system. The conventional design creates a drum about 4 feet (1.2 m) long almost in contact with the ground. In squeezing ground this shield is vulnerable to the pressure exerted by rock movement. It is therefore better that the shield be smaller in diametre than the excavation and that it be tapered toward the rear. The gauge cutters should be arranged to protrude beyond the main body of the cutterhead.

If the cutterhead is not in close contact with the ground, provision must be made to provide stable support in its place. This will be the equivalent of a sole plate as used for overcutter compensation in earth pressure balance machines. However, in order to provide for varying amounts of overcut, the support will need to be hydraulically actuated. Since it will be subjected to substantial shear loading, the design will have to be very stiff.

Propulsion

A TBM requires a reaction against which to propel itself forward. This reaction can be obtained by shoving directly against the tunnel support system with jacks spaced around the perimetre of the machine or by developing frictional resistance against the tunnel sidewalls.

The thrust needed to keep the cutterhead moving forward is about 25,000 kg per cutter. When the ground is weak, it is desirable to limit the bearing pressure on the tunnel walls because the weak rock would fail under even light loads, especially perpendicular to the direction of foliation. This would accelerate the rate of squeezing and might increase the total strain. At the same time it would be desirable to limit the length occupied by the grippers so as to minimise the necessary distance between the working face and any support system. This would probably require that there be multiple grippers covering most of the circumference but of limited length to minimise uneven bearing on the squeezing rock surface.

Shield

If any shield is felt to be desirable or necessary, it should be short and shrinkable. Many TBMs have been stuck because the ground has moved on to the shield and exerted sufficient load to stall the machine.

Erector

It is desirable to have complete flexibility in selecting the point at which ring erection is to take place. Therefore the erector should be free to move along the tunnel, mounted on the conveyor truss. A ring former should also be used to maintain the shape of the last erected ring until it has been grouted if concrete segmental lining is used.

Spoil Removal

Conventional conveyor to rail car systems or single conveyor systems designed for the tunnel size selected are appropriate.

Back-Up System

In order to keep the area between the grippers and the ring erection area as clear as possible, any ancillary equipment such as transformers, hydraulic pumps etc. should be kept clear of this space at track level.

Operational Flexibility

It is envisaged that the system outlined above would be capable of handling either steel ribs or precast concrete supports. If shoving off the supports were to be selected for TBM propulsion, the degree of flexibility would be less than with the use of a gripper system. It would also be more vulnerable to problems in any circumstance where the convergence rate was markedly higher than expected.

Swelling

Swelling phenomena are generally associated with argillaceous soils or rocks derived from such soils. In the field, it is difficult to distinguish between squeezing and swelling ground, especially since both conditions are often present at the same time. However, except in extreme conditions, squeezing is almost always self-limiting and will not recur vigorously, or at all, once the intruding material has been removed; while swelling may continue as long as free water and swelling minerals are present especially when the intruding material has been removed, thereby exposing fresh, unhydrated rock. Many European rail and highway tunnels are constructed in formations noted for their susceptibility to swelling. Most construction involves a more or less circular wall and roof section with an invert slab having a greater radius of curvature. Some of them are still being periodically repaired a century after construction. It has been noted in this connection that as the invert arches are excavated and replaced to more nearly circular configurations, the greater the time that elapses before the next repair is necessary.

Expansive clays are more common in younger argillaceous rocks, the proportions ranging from 65% in Pliocene and Miocene age material to only 5% in Cambrian and Precambrian. Montmorillonite is found in rocks of all ages as thin partings or thicker beds. Sodium montmorillonite is much more expansive than calcium montmorillonite.

Swelling Mechanism

Most swelling is due to the simultaneous presence of unhydrated swelling clay minerals and free water. Tunnel construction commonly creates these conditions. Minerals such as montmorillonite form layered platy crystals; water may be taken up in the crystal lattice with a resultant increase in volume of up to ten times the volume of the unhydrated crystal. The displacements resulting from this increase in volume give rise to the observed swelling pressures, whether in soil or in rock.

If possible water should be kept away from rock or soil containing swelling clay minerals; however, it must be realised that water from fresh concrete, water vapour from a humid atmosphere or pore water released from confinement within the rock will initiate the swelling process. Since the swelling will not passivate in the same way as squeezing generally will in rock, tunnel support must be designed to resist the swelling pressure (which can be measured in the laboratory), even if it proves possible to let some swelling take place without creating problems.

Other Rock Problems

Schists commonly contain clay minerals such as biotite, mica and chlorite. All of these are platy minerals and are found aligned with the foliation. If present as continuous layers, they have to be considered planes of weakness when assessing questions of rock stability. Similarly, weathered material in shears and mylonite not yet weathered indicate planes of weakness.

Anhydrite converts to gypsum in the presence of water with a volume increase of up to 60% accompanying the conversion. However, beds of anhydrite are not affected in the same way as finely divided rock since the reaction does not penetrate below the surface. However, if the anhydrite is fractured, the conversion will proceed faster and faster as more fracturing is developed by the expansive reaction. The actual amount of expansion will depend upon the void ratio of the anhydrite. As with other water-sensitive minerals, every effort should be made to keep water away from anhydrite. This may be a particular problem when fluid transport tunnels are being constructed since any leakage will result in major damage to the tunnel.

Obstacles and Constraints

Boulders

Practical experience of the value of cutterhead disks in such a situation was first developed in Warrington, England . A slurry shield was to be used for a tunnel originally expected to be in soils. A late decision to change the alignment because of local constraints forced the tunnel into an area where boulders and sandstone bedrock would be encountered in the invert. Since the equipment was already built, disk cutters were added to the head in the hope that they would solve the unexpected problem. These hopes were fulfilled. More recent investigation in Japan has indicated from experimental models that even very soft clay will provide sufficient support to hold boulders in place so that they are broken up by the action of disk cutters. On the other hand, rotary head excavators of various general designs have failed to deal successfully with boulders when drag picks were relied on.

A particular difficulty sometimes occurs when boulder beds are encountered which have saturated fine silt in the void spaces between the boulders. This problem seems to be most often encountered in regions which have been subjected to glaciation. The loss of ground associated with flow of the saturated fines into the tunnel does not normally result in ground settlement, because the movement of any other material replacing the lost fines will generally be choked off. If this is not the case, or if it is felt undesirable to leave such voids unfilled, various courses of action are available. Compressed air working will drive water out of the silt and thereby stabilise it, provided that the boulder bed is not confined within impervious material. In such a case, compressed air working will not be very effective.

The use of an EPB fitted with disk cutters will be effective provided that the pressure in the plenum chamber is kept at a level high enough to balance the hydrostatic head in the silt. Slurry shield operation with the same restrictions would be even more effective, but at a higher cost. As a last resort, consolidation or replacement grouting may be employed behind the shield. The choice of method will depend on economics, as is often the case when selecting a construction method. If the condition exists in only a small part of a long tunnel, less efficient means may be selected for dealing with the boulder bed—even including local cut-and-cover work, if the tunnel is not too deep or the water table too high. In any case, full breasting of the face is required if the boulders are not in intimate contact with one another. It is conceivable that grout could be injected into the working face at a distance behind it so as to force out the flowing material. For such a program to be effective, it would be necessary to grout multiple points simultaneously so as to avoid development of a preferred path for escaping fines. The grout would also have to extend outside the tunnel perimetre for a sufficient distance to establish a plug which could be excavated without developing problems behind the shield. However, it must be said that in small tunnels, access for implementation of such a program is unlikely to be available.

Karstic Limestone

Karstic limestone is often riddled with solution cavities of various sizes. Depending on the geologic history of the locale in which it is found, cavities ahead of the excavation may be filled with water, mud or gravel or a combination of these. Flowing water may be present in large quantities. There may be an insufficient thickness of sound rock at tunnel elevation to provide safe support for tunnelling equipment. All of these possibilities point out the need for thorough exploration before undertaking tunnel construction in limestone, particularly in an area where there is no prior history of underground construction or mining.

Abandoned Foundations

Abandoned foundations or other facilities, are to be expected in urban tunnels. To illustrate, on one project 898 piles were encountered during construction of rapid transit tunnels in an urban area. This was more than double the highest estimate. These piles were mostly unrecorded relics of earlier construction abandoned after fires which regularly ravaged the area during the late 19th century as well as piles left behind by successive reclamation operations, which moved the waterfront several hundred metres into the bay over a few decades.

All but two of the piles were timber; they were removed by cutting them into short lengths as they were exposed in the face of the shield using a hydraulically powered beaver-tail chain saw purpose-made for the job. The other two posed a different problem. One was concrete and the other steel.

Since the tunnel was being constructed in compressed air, both burning the steel and breaking the concrete were non-trivial problems. Removal of these piles took about 10 times as long as removal of the timber piles. With a TBM, similarly, the wood piles can be cut by the disk cutters but a similar increase in time would be expected for the steel and concrete piles.

As an additional problem, the lengths of pile left in place above the tunnel eventually crept downward as they sought to carry the weight of soil adhering to them as well as the artificial fill above. In a number of places it was necessary to reinforce the skin of the fabricated steel liner plates which were dimpled by the point loads exerted.

A second illustrative project involved construction of a storm drainage tunnel. For a short distance at the downstream portal, the tunnel was in soil; because of its short length it was driven without a shield. Since the soil was largely, if not entirely, fill, it proved difficult to maintain the tunnel shape until steel ribs were introduced between alternate rings of liner plate. It was known that old mill foundations lay ahead, but their location was uncertain. It was therefore deemed prudent to continue this tunnelling method into the sandstone ahead for at least a short distance.

During the drive, some of the foundations were found in the soil tunnel. Careful breasting to isolate the concrete was successful in controlling soil movement while the concrete was broken out. With the next advance of normal tunnelling, the voids were promptly and completely filled and there was no encroachment on the tunnel profile.

It would be possible to multiply examples endlessly, but the key to all such problems is to gather the maximum available information, project the worst scenario and be prepared to deal with it as an engineering rather than an economic problem.

Shallow Tunnels

The problem with shallow tunnels is that side support is not reliable and loading on the support system is far from the usual comfortable assumption of essentially uniform radial load. It is quite common in urban situations to be restricted by the presence of significant structures—whether on the surface or underground.

Consolidation grouting has been used where ground conditions are favourable and compaction grouting has also been used successfully to avoid the need for underpinning. In other cases; jet grouting, pipe canopies or other spiling may be appropriate. It is not practical to define the range of conditions leading to selection of any particular solution, since all such projects tend to have unique features. Sequential excavation method, cut and cover method and/or jacked box tunnelling method can be considered as well.

Chapter 6

Physical Conditions

Methane

Methane is commonly found where organic matter has been trapped below or within sedimentary deposits, whether or not they have yet been lithified. It is particularly common in the shaley limestones around the Great Lakes; in hydrocarbon—whether coal or oil— deposits in Pennsylvania, West Virginia, Colorado or California and in many other localities. It should be noted here that "methane" is commonly used as a denotation of all of the ethane series that may be present, although methane is distinguished as the major component usually present. It is also the only member substantially lighter than air. Methane forms an explosive mixture when mixed with air and between about 5 and 15% of the total volume is methane. It is readily diluted and flushed from a tunnel by ventilation when encountered in the quantities that are normally expected.

Safety rules require that action be taken when methane is present in concentrations of 20% of the lower explosive limit. For practical purposes, this means a concentration of 1% by volume. It is necessary to use routine testing to determine whether or not explosive gasses are present. This testing is carried out in all tunnels identified as being gassy or potentially gassy. A positive rating of non-gassy is required to relieve the contractor of the duty to test, although in many localities it is deemed prudent to continue testing on a reduced schedule even though no gas has been identified in the tunnel excavation.

Hydrogen Sulfide

Hydrogen sulfide is present in association with methane often enough that its presence should always be suspected in gassy conditions. Its presence is easily identified in low concentrations by its typical rotten egg smell. It is a cumulative poison and deadly in low concentrations; a whiff at 100 percent concentration is generally instantly fatal. It should also be noted that, when hydrogen sulfide is present in low concentrations, the nose becomes desensitized to its presence. Apart from testing and maintenance of high volume ventilation, signs of its presence follow a sequence of headaches, coughing, nausea and unconsciousness. Concentrations should be limited to 10 ppm or less (depending on local regulations) of eight hour exposures.

Meticulous attention to ventilation, especially in work areas, is required when hydrogen sulfide is present or suspected. As with methane, ventilation must be maintained at high volumes for dilution 24 hours a day, 7 days a week, regardless of whether work is in progress or not. Even so, no shaft, pit or tunnel or other opening below grade should be entered without first testing the air. This is especially important if the purpose of entering is to repair a defective fan. Where possible, the gas should be extracted directly and discharged into the ventilation system without ever entering the tunnel atmosphere.

High Temperatures

The geothermal gradient is different in different localities within a range of about 2:1. As a rule of thumb, one degree Celsius per 100 metres of depth will be a reasonable guide. Where the tunnel is comparatively shallow—say less than about 150 m—there will be little effect. In fact, it will be found that the tunnel temperature is the average year-round temperature at that location.

Nevertheless, especially in areas of volcanism or geothermal activity or tropical temperatures, the temperature in deep tunnels can rise to body heat or higher. If hot water flows are present or if the tunnel is very humid (which is more common than not), conditions can be actively dangerous as sweating and evaporation are inhibited; heat stroke can be induced in such conditions. The only factor that can be directly controlled is the tunnel ventilation. By supplying air at a lower temperature, the local conditions can be kept bearable, especially if the incoming air is dry enough to accept evaporating moisture.

Observations

The significant effects and the construction problems resulting from the various difficult tunnelling conditions discussed above make it clear that all of the possibilities associated with the geology and occupational history of the region in which new tunnelling is contemplated need to be borne in mind from the construction as well as the design standpoint when the preliminary and final geotechnical exploration and testing programs are designed.

It is important that engineers designing a tunnel project develop a full understanding of the nature of the ground conditions affecting the construction; so that not only the field investigation but also the design development, specifications and geotechnical reports reflect a full understanding of the problems and the variety of potential approaches to their solution. In the end, the project owner's interests will be best served by thoughtful analysis and full disclosure of conditions and of the solutions foreseen, and of the underlying design approach rather than by avoiding the recognition of problems and their potential impact.

Sequential Excavation Method (SEM)

The Sequential Excavation Method (SEM), also commonly referred to as the New Austrian Tunnelling Method (NATM), is a concept that is based on the understanding of the behaviour of the ground as it reacts to the creation of an underground opening. In its classic form the SEM/NATM attempts to mobilise the self-supporting capability of the ground to an optimum thus achieving economy in ground support. Building on this idea practical risk management and safety requirements add to and dictate the required tunnel support. Initially formulated for application in rock tunnelling in the early 1960's, NATM has found application in soft ground in urban tunnelling in the late 60's and has since then enjoyed a broad, international utilisation in both rural and urban settings.

A large number of tunnels have been built around the world using a construction approach which was loosely termed NATM. During the years of discussions and the application of NATM a variety of terms have been used for the same construction approach. These terms were primarily aimed at describing the construction approach rather than the region of its reported origin. While in the 70's and early 80's the term "Shotcrete Method" was frequently used in Germany and Switzerland, besides NATM, developments in the UK in the late 90's led to the use of the term "Sprayed Concrete Lining" or SCL. Alternatively, "Conventional Tunnelling Method" was used in Austria and Germany . As the NATM is largely based on an observational approach, the term "Observational Method" was introduced and used in many countries. The term "Conventional" as opposed to TBM driven tunnels has recently found its way into publications by the International Tunnelling Association's (ITA) Working Group 19. In the German speaking countries in Europe namely in Austria and Germany very recent standards and codes use the term "Cyclic Tunnelling Method."

In the US, where NATM was systematically applied for the first time in the late 70's and early 80's for the construction of the Mount Lebanon tunnel in Pittsburgh and the Redline tunnels and Wheaton Station of the Washington, DC metro the term adopted was NATM. Gradually, however, the term has been and is being abandoned in the US and replaced by Sequential Excavation Method or SEM. Today, SEM is becoming increasingly popular in the US for the construction of tunnels, cross passages, stations, shafts and other underground structures (Gildner et. al., 2004).

The SEM offers flexibility in geometry such that it can accommodate almost any size of opening. The regular cross section involves generally an ovoid shape to promote smooth stress redistribution in the ground around the newly created opening. By adjusting the construction sequence expressed mainly in round length, timing of support installation and type of support it allows for tunnelling through rock, soft ground and a variety of difficult ground conditions. Depending on the size of the opening and quality of the ground a tunnel cross section may be subdivided into multiple drifts.

Application of the SEM involves practical experience, earth and engineering sciences and skilled execution. The SEM tunnelling process addresses:

- Ground and excavation and support classification based on a thorough ground investigation
- Definition of excavation and support classes by:
  - Round length (maximum unsupported excavation length)
  - Support measures (shotcrete lining and its reinforcement, ground reinforcement by bolts or dowels in rock)
  - Subdivision of the tunnel cross section into multiple drifts or headings as needed (top heading, bench, invert, side wall drifts)
  - Ring closure requirements
  - Timing of support installation (typically every round)
  - Pre-support by spiling, fore poling, and pipe arch canopy
  - Local, additional initial support by dowels, bolts, spiles, face support wedge, and shotcrete

- Instrumentation and monitoring
- Ground improvement measures prior to tunnelling
- A key support element is shotcrete mainly due to its capability to provide an interlocking, continuous support to the ground. Implementation of ground improvement measures in the form of dewatering, grouting, ground freezing and others and of pre-support measures in the various forms of spiling have further widened the range of SEM applications mainly in urban tunnelling. These are specified to enhance the quality of the ground prior to and during the tunnelling process. The SEM features typically a dual lining cross section by which a waterproofing membrane is inserted between an initial shotcrete and a final, typically cast-in-place concrete lining. Tunnel Lining. An instrumental element of SEM tunnelling is the monitoring of deformations of tunnel and surrounding ground. Evaluation of monitoring allows for the verification of design assumptions or adjustment of the tunnelling process.
- Lastly, because SEM tunnelling allows for an adjustment to ground conditions as encountered in the field it benefits from a unit-price contract form. Geotechnical baseline reports (GBR) facilitate the adjustment process and aid in risk sharing between owner and contractor.

Background and Concepts

The origins of the NATM lie in the alpine tunnel engineering in the early 1960s. In 1948, Ladislaus von Rabcewicz applied for a patent for the use of a dual lining system with the initial lining being allowed to deform. The NATM is based on the philosophy that the ground surrounding the tunnel is used as an integrated part of the tunnel support system. The deformable shotcrete initial lining allows a controlled ground deflection to mobilise the inherent shear strength in the ground and to initiate load redistribution. The key for the successful use of a relatively thin lining layer applied to the excavation surface lies in the smooth tunnel shape to avoid stress concentrations and the tight contact between the shotcrete lining and the surrounding ground to provide an intense interaction between the support and the ground. In order to augment the support provided by the initial lining, rock reinforcement is used in response to the rock mass conditions. The rock reinforcement avoids the development of wedge failure (keystone), and it generates a rock mass ring with significantly improved strength characteristics around the opening. Smooth, concavely rounded excavation surfaces initiate confinement forces and limit bending and tension forces in the lining and the ground in the vicinity of the tunnel opening. This is of particular importance for tunnelling in ground with limited stand-up time, where fracturing and weathering reduce the ground's natural shear strength.

NATM was the first concept, where the ground and its strength were used as a building material and became an integrated part of the tunnel support system. Rather than implementing stiff support members that attract high loads to fight the ground deformation, the flexible, yet strong shotcrete lining shares with and re-distributes loads into the ground by its deflection.

Rabcewicz summarises the philosophy of NATM in his patent of 1948 (Rabcewicz, 1948) as follows: "NATM is based on the principle that it is desirable to take utmost advantage of the capacity of the rock to support itself, by carefully and deliberately controlling the forces in the readjustment process which takes place in the surrounding rock after a cavity has been made, and to adapt the chosen support accordingly." By briefly reviewing the stress conditions around a newly created cavity and the interaction between ground and its support needs the following lays out the principle approach taken in NATM tunnel design (Rabcewicz et al., 1973).

The stress conditions around a cavity after Kastner are schematically provided. The primary stress σ_0 in the surrounding ground before any cavity is created depends mainly on the overburden, the unit weight and any tectonic stresses σ_s. Following tunnel excavation the tangential stresses will increase next to the tunnel circumference (solid line σ_t^0). If the induced tangential and radial stresses (σ_t and σ_r) around the tunnel opening exceed the strength of the surrounding ground yielding will occur. Such yielding will create a plastified zone that reaches to a certain distance R into the ground beyond the tunnel circumference (dashed line R).

The relationships between the radial stresses σ_r, deformation of the tunnel opening Δr, the required outer and inner supports p_i^a and p_i^I respectively, and the time of support application T is provided. According to Rabcewicz the outer support or outer arch (p_i^a) involves the ground itself, its reinforcement by rock bolts and any support applied to the opening itself ranging from sealing shotcrete (flashcrete) to a structural initial lining involving reinforced shotcrete or concrete and steel ribs. The inner support involves a secondary lining that is applied after the tunnel opening with the help of the outer arch has reached equilibrium.

The σ_r /Δr curve, often referred to as ground reaction curve schematically describes the relationship between deformation of the tunnel opening and tunnel support provided by the outer arch. At any intersection point between the support p_i and the σ_r curve equilibrium is reached for the respective support. It is characteristic for the NATM that the intersection between the support and the σ_r curve takes place at the descending side of the curve. Undesirable loosening of the ground starts at point B of the σ_r curve if the minimum support p_i min is not provided. Within the ascending side of the σ_r curve the ground has lost strength and consequently its supporting capacity and thus requires enhanced tunnel support to passively support the overburden.

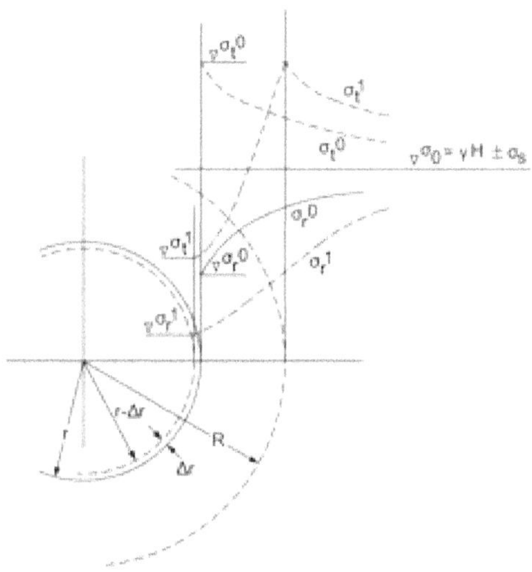

Figure: Schematic Representation of Stresses Around Tunnel Opening

Examination of curves the relationship between timing of support installation, yielding of the ground and the amount of support needed. The minimum support is required at point B to prevent loosening and loss of strength in the surrounding ground. It will result in the largest deformations but the most economical tunnel support. Curve 1 which intersect the σ_r curve in point A will require enhanced support p_i^a but result in less deformation Δr and a higher factor of safety. Selection of a stiffer outer arch in curve 2 will result in more support loads because the ground has not been allowed to deform and mobilise its strength and consequently led to a decrease of the safety factor.

The capacity of the inner arch is chosen to satisfy a desired safety factor s. This will depend on specific needs and assuming that the initial tunnel support (outer arch) will deteriorate over time then p_i^a may be used as guidance to arrive at a desired safety factor. C and C' denote a loaded and unloaded condition of the inner arch respectively.

The σ_r /Δr curve may be approximated by means of numerical modelling using the deformation and strength characteristics of the ground along with the specific geometry of the opening and the envisioned excavation sequencing.

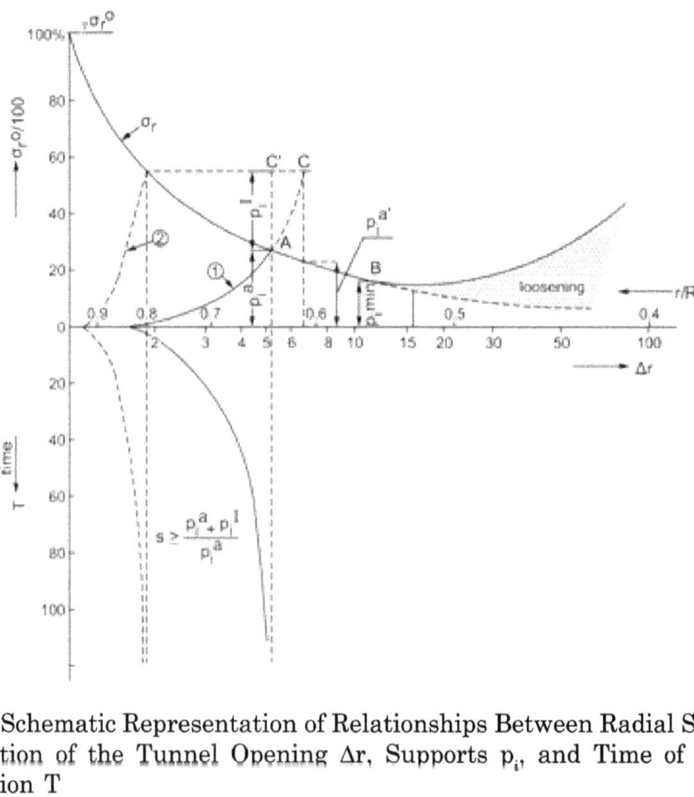

Figure: Schematic Representation of Relationships Between Radial Stress σ_r, Deformation of the Tunnel Opening Δr, Supports p_i, and Time of Support Installation T

While the NATM had its origins in alpine tunnelling in fractured or squeezing rock, its field of application expanded dramatically in the 70's and the following decades. The superb flexibility of the construction concept to adapt to a wide range of ground conditions and tunnel shapes in combination with significant developments in construction materials, installation techniques as well as ground treatment methods formed the basis for a radical expansion from the alpine rock tunnelling into soft ground tunnelling. The use of NATM thus expanded from rural tunnelling in rock into urban tunnelling in predominantly soft ground and highly built-up environments with sensitive structures above the tunnel alignment.

The major focus in rock tunnelling in rural settings is to find equilibrium in the surrounding ground with the best possible economy in the amount of initial support installed. In urban settings however, in particular when tunnelling at shallow overburden depths in soft ground the main goal is to minimise the impact on the surface and adjacent structures thus to minimise settlements. As shown, less and delayed support installation will be associated with larger deformations of the tunnel Δr and consequently with larger surface settlements when tunnelling at shallow depth. Curve No. 1 describes a relatively "soft" support that is applied later than the support represented by curve No. 2, which is applied earlier and is "stiffer." The curves point out that in order to reduce settlements generally an early and stiffer support should be used. Reduction of the round length and subdivision of the tunnel cross section will aid in applying support to the ground early thus reducing deformations. The stiffness of the support can be increased by increasing the initial shotcrete lining thickness and using shotcrete with early and high strength development.

Today's tunnel construction economies require tunnelling approaches that are competitive to fully mechanized tunnelling methods by TBMs with their high initial capital cost while being adjustable to project specific space demands. The main field of SEM application is, apart from rural railway and highway tunnels, in the construction of tunnel schemes with complex geometries, short tunnels, large size tunnels and caverns in urban areas at shallow depths. Shallow tunnel depths frequently involve the challenge of soft ground tunnelling. With the help of modern equipment for rapid excavation, modern high quality construction materials (mainly shotcrete), and modern ground support installation techniques as well as the overarching SEM concept, complex and challenging underground structures can be built in practically all types of ground. A major advantage of the SEM is its flexibility.

SEM Regular Cross Section

Geometry

The shape of the tunnel cross section is designed to comply with SEM principles, which are to (as effectively as possible) activate the self-supporting arch in the surrounding ground. To accommodate this principle cross section geometries shall be curvilinear, consisting of compound curves in both arch and invert (if constructed in soft ground like conditions). Any straight walls and sharp edges in the excavation cross section shall be avoided. Thus the geometry of the excavation cross section will enable a smooth flow of stresses in the ground around the opening, minimising loads acting on the tunnel linings. While adhering to these principles the excavation cross section shall be optimized in size to achieve economy. The layout of the invert will depend on the ground conditions in which the tunnel is constructed. In competent rock formations the tunnel invert will be flat, whereas in weak rock and soft ground tunnels the invert will be rounded to facilitate ring closure and stability.

Dual Lining

The SEM regular cross section is of dual lining character and consists of an initial shotcrete lining and a final, cast-in-place concrete or shotcrete lining. A waterproofing system is sandwiched between the initial and final linings. The waterproofing system consists of a flexible, continuous membrane (typically PVC). A regular cross section is developed for each tunnel geometry: the main tunnel, widenings, niches, cross passages, and other miscellaneous structures. A typical regular SEM cross-section for a two-lane highway tunnel is shown, distinguishing between a rounded (right side) and flat (left side) invert. A rounded invert is typically associated with tunnelling in soft ground whereas a flat invert is used in competent ground conditions, typically rock. The tunnel cross section is designed around the project clearance envelope including tolerances. The alignment of the tunnel is curved to accommodate alignment needs of an urban environment. In the front the SEM tunnel abuts a straight tunnel wall of an adjoining cut-and-cover box tunnel.

Figure: Three-Lane SEM Road Tunnel Interior Configuration (Fort Canning Tunnel, Singapore)

Initial Shotcrete Lining

The initial shotcrete lining is the layer of shotcrete applied to support the ground following excavation. It has a thickness ranging generally from 4 to 16 inches (100 to 400 mm) mainly depending on the ground conditions and size of the tunnel opening. It is reinforced by either welded wire fabric or steel fibers; the latter have generally replaced the traditional welded wire fabric over the last ten to fifteen years.

Occasionally structural plastic fibers are used in lieu of steel fibers. This is the case where the shotcrete lining is expected to undergo high deformations and ductility post cracking is of importance. Where the shotcrete lining is greater than about 6 inches (150 mm) it further includes lattice girders. Depending on loading conditions and purpose rolled steel sets may replace lattice girders or act in combination.

Waterproofing

The SEM uses flexible, continuous membranes for tunnel waterproofing. Most frequently PVC membranes are used at thicknesses of 80 to 120 mil (2.0 to 3.0 mm) depending on the size of the tunnel. Only in special circumstances, for example when contaminated ground water is present, special membranes are applied using hydrocarbon resistant polyolefin or very light density polyethylene (VLDPE) membranes.

The impermeable membrane is backed by a geotextile that also acts as a protection layer, and in drained systems as a drainage layer behind the membrane. This waterproofing system is placed against the initial lining and prior to installation of the final lining. Prior to waterproofing system installation all tunnel deformations must have ceased.

In drained system applications water is collected behind the membrane and conducted to perforated sidewall drainage pipes located at tunnel invert elevation on each side of the tunnel. From there collected water is conveyed via transverse, non-perforated pipes to the tunnel's main roadway drain. In undrained systems the membrane and geotextile wrap around the entire tunnel envelope and prevent water seepage into the tunnel thereby subjecting it to hydrostatic pressures. If this is the case the tunnel invert geometry and structural design must be adapted to accommodate for the hydrostatic head.

Over the past decades a so called "compartmentalization system" has been developed and nowadays supplements the installation of flexible membrane based waterproofing systems. The purpose of this compartmentalization is to provide repair capability in case of leakage. In particular, when the tunnel is not drained and the waterproofing has to withstand long-term hydrostatic pressures, installation of such systems provides a cost effective back up and assures a dry tunnel interior. Compartmentalization refers to the concept of subdividing the waterproofing membrane into individual areas of self-contained grids (compartments) by means of base seal water barriers. These water barriers are specifically formulated for the purpose of creating these compartments. They feature ribs of 1.3-inch (30 mm) minimum height to properly key into the final lining, which is cast (or sprayed) against the waterproofing.

In case of water leakage the water infiltration is limited to the individual compartment thus preventing uncontrolled water migration over long distances behind the final lining. Within each compartment control and grouting pipes are installed. These pipes penetrate through the final lining and are in contact with the membrane. Control and grouting pipes serve a twofold purpose; should leakage occur then water would find its path to these pipes and exit there thus signalling a breach within the compartment. Once detected, the same pipes may be used for injection of low viscosity, typically hydro-active grouts into the compartments. The injection of grout is limited to leaking compartment(s) and once cured provides a secondary waterproofing layer in the form of a membrane that acts as a remedial waterproofing layer.

Figure: Waterproofing System and Compartmentalization (Automated People Mover System at Dulles International Airport, Virginia)

Smoothness Criteria

To provide a suitable surface for the installation of the waterproofing system, all shotcrete surfaces to which the membrane is to be applied must meet certain smoothness criteria. These are expressed in the waviness of the shotcrete surface to which the waterproofing system will be applied. The waviness is measured with a straight edge laid on the surface in the longitudinal direction. The maximum depth to wavelength ratio should be generally 1:5 or smoother. The surface has to be inspected prior to installation of the waterproofing system and all projections should be removed or covered by an additional plain shotcrete layer, which meets the smoothness criteria. The SEM design documents will address required smoothness criteria and set those in relation to the waterproofing system to be used.

Final Tunnel Lining

The final permanent lining for a SEM tunnel may consist of cast-in-place concrete or shotcrete. Cast-in-place concrete can be unreinforced or reinforced. Shotcrete is generally fibre reinforced. The following addresses design and construction considerations specifically for SEM application.

Cast-in-Place Concrete Final Lining

The traditional final lining consists of cast-in-place concrete at a thickness of generally 12 inches for two-lane road tunnels. While the lining may generally remain unreinforced, structural design considerations and project design criteria will dictate the need for and amount of reinforcement. The Lehigh Tunnel (Pennsylvania) and Cumberland Gap Tunnels (Kentucky / Tennessee) are the first road tunnels built in the US in the late 80's and early 90's using SEM construction methods. Both feature unreinforced, 12-inch thick cast-in-place concrete final linings. The flexible membrane based waterproofing is in particular beneficial in unreinforced cast-in-place concrete lining applications in that it acts as a de-bonding layer between the initial and final linings and therefore reduces shrinkage cracking in the final lining.

To ensure a contact between the initial and final linings, contact grouting is performed as early as the final lining has achieved its 28-day design strength. With this grouting the contact is established between the initial lining and final tunnel support. Any deterioration or weakening of the initial support will lead to an increased loading of the final support by the increment not being supported by the initial lining. The loads can be directly transferred radially due to the direct contact between initial and final linings. Cast-in-place final concrete linings (concrete arch placed on sidewall footings) are frequently installed in pour lengths not exceeding 30 feet (10 metres). This restriction is important to limit surface cracking in general and becomes mandatory if unreinforced concrete linings are used. A 30 feet (10 metre) long section in a typical two-lane highway tunnel is also practical in terms of formwork installation and sequencing and duration of concrete placement.

Adjacent concrete pours feature construction joints that are true lining separators designed as contraction joints. The inside face at joint location shall be laid out with a trapezoidally shaped joint. A continuous reinforcement is not desired in construction joints to allow their relative movement in particular for thermal deformation effects.

Water Impermeable Concrete Final Lining

Use of water impermeable cast-in-place concrete linings as an alternative to membranes is generally not considered due to the high demands on construction quality and exposure to freeze thaw conditions in cold climates. Elaborate measures are needed to prevent cracking. Detailed arrangement of construction joints is needed as well as complex concrete mix designs to suppress excessive hydration heat. The curing requires elaborate procedures. These aspects generally do not render water impermeable concrete practical in road tunnels. If selected these construction aspects have to be addressed in detail in specifications and working procedures and they have to be rigidly enforced.

Shotcrete Final Lining

Shotcrete represents a structurally and qualitatively equal alternative to cast-in-place concrete linings. When shotcrete is utilised as a final lining in dual lining applications it will be applied against a waterproofing membrane. The lining thickness will be generally 12 inches (300 mm) or more and its application must be carried out in layers with a time lag between layer applications to allow for shotcrete setting and hardening. Its surface appearance can be tailored to the desired project goals. It may remain of a rough, sprayer type shotcrete finish, but may have a quality comparable to cast concrete when trowel finish is specified. Shotcrete as a final lining is typically utilised when the following conditions are encountered:

- The tunnels are relatively short in length and the cross section is relatively large and therefore investment in formwork is not warranted, i.e. tunnels of less than 100-250 m (300-800 feet) in length and larger than about 8-12 m (25-40 feet) in springline diametre.

- The access is difficult and staging of formwork installation and concrete delivery is problematic.
- The tunnel geometry is complex and customized formwork would be required. Tunnel intersections, as well as bifurcations qualify in this area. Bifurcations are associated with tunnel widenings and would otherwise be constructed in the form of a stepped lining configuration and increase cost of excavated material.

Figure displays a typical shotcrete final lining section with waterproofing system, welded wire fabric (WWF), lattice girder, grouting hoses for contact grouting and a final shotcrete layer with PP fibre addition.

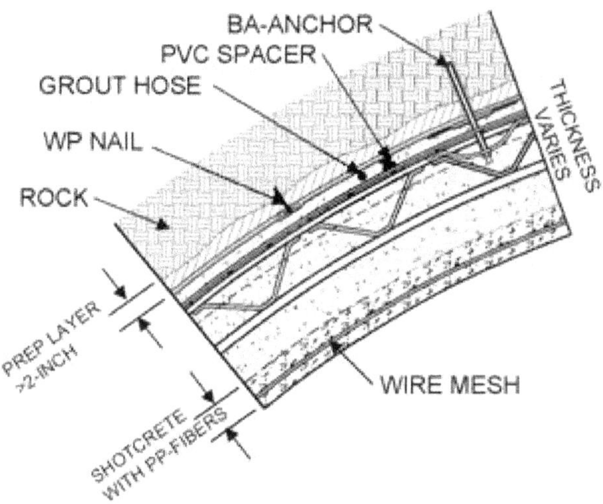

Figure: Typical Shotcrete Final Lining Detail

Single Pass Linings

Under special circumstances the initial shotcrete lining alone or with the addition of an additional shotcrete layer designed to withstand long-term loads may be used as a single support lining for the long term. Although labelled "single pass" this final shotcrete lining may be applied in multiple shotcrete application cycles. Use of a single pass lining will generally be limited to conditions where the ground water inflow is not of concern and deterioration of the shotcrete product over the life time of the tunnel lining can be excluded or partially tolerated.

In multiple layer applications the shotcrete surface to which additional layers will be applied must be sufficiently clean and free of any layer that may cause de-bonding over the long term (Kupfer, et al., 1990 and Hahn, 1999). Specially detailed construction joints and high quality shotcrete must be required to assure water tightness and long-term integrity.

Ground Classification and SEM Excavation and Support Classes

Rock Mass Classification Systems

A series of qualitative and quantitative rock mass classification systems have been developed over the years and are implemented on tunnelling projects worldwide. An overview of the most commonly used rock mass classification systems including Terzaghi's qualitative classification, and quantitative systems such as the Q system and the Rock Mass Rating (RMR) system.

Rock mass classification systems aid in the assessment of the ground behaviour and ultimately lead to the definition of the support required to stabilise the tunnel opening. While the above quantitative classification systems lead to a numerical rating system that results in suggestions for tunnel support requirements, these systems cannot replace a thorough design of the excavation and support system by experienced tunnel engineers.

Ground Support Systems

In the early years of the use of NATM (SEM) in Austria, Switzerland and Germany, standards and codes used descriptive (qualitative) categories to define ground support classes. Recent standards, codes and guidelines implemented in Austria and Germany utilise a process-oriented approach (OGG, 2007). This approach defines the process of using relevant parameters from ground investigation to derive a ground response classification and subsequently assess tunnel support needs. This forms a more objective basis for all parties involved and promotes the understanding of the rationale in retrospect by persons that have not been involved in the design process. It also provides a common platform for contractors, owners and engineers to negotiate the project specific challenges in the field during actual construction.

All classification systems have in common that they should be based on thorough ground investigation and observation. The process from the ground investigation to the final definition of the ground support system can be summarised in three models:

- Geological Model
- Geotechnical Model
- Tunnel Support Model

Geological Model

A desk study of the geological information available for a project area forms the starting point of the ground investigation program. Literature, maps and reports (e.g. from the US Geological Survey) form the basis for a desk study. Subsequently and in coordination with initial field mapping results, a ground investigation program is developed and carried out. The geological information from the ground investigation, field mapping, and the desk study are compiled in the geological model.

Geotechnical Model

With the data from the geological model in combination with the test results from the ground investigation program and laboratory testing the ground response to tunnelling is assessed. This assessment takes into account the method of excavation, tunnel size and shape as well as other parameters such as overburden height, environmental issues and groundwater conditions. The geotechnical model assists in deriving zones of similar ground response to tunnelling along the alignment and Ground Response Classes (GRC) are defined. These GRCs form the baseline for the anticipated ground conditions. Typically, the ground response to an unsupported tunnel excavation is analysed in order to assess the support requirements for the stabilisation of the opening (OGG, 2007).

Tunnel Support Model

After assessing the ground support needs, excavation and support sequences, subdivision into multiple drifts, as well as the support measures are defined. These are combined in Excavation and Support Classes (ESCs) that form the basis for the Contractor to develop a financial and schedule bid as well as to execute SEM tunnel work.

Excavation and Support Classes (ESC) and Initial Support

Excavation and Support Classes (ESCs) contain clear specifications for excavation round length, subdivision into multiple drifts, initial support and pre-support measures to be installed and the sequence of excavation and support installation. They also define means of additional initial support or local support or pre-support measures that augment the ESC to deal with local ground conditions that may require additional support. In SEM tunnelling initial support is provided early on. In soft ground and weak rock it directly follows the excavation of a round length and is installed prior to proceeding to the excavation of the next round in sequence. In hard rock tunnelling initial support is installed close to the face. The intent is to provide structural support to the newly created opening and ensure safe tunnelling conditions. Initial support layout is dictated by engineering principles, economic considerations, and risk management needs.

The amount and design of the initial support was historically motivated mainly by the desire to mobilise a high degree of ground self support and therefore economy. This was possible at the outset of SEM applications in "green field" conditions where deformation control was of a secondary importance and tolerable as long as equilibrium was reached. Nowadays, however, safety considerations, risk management, conservatism and design life, and the need for minimising settlements in urban settings add construction realities that ultimately decide on the layout of the initial support.

Initial support is provided by application of a layer of shotcrete to achieve an interlocking support with the ground. Shotcrete is typically reinforced by steel fibers or welded wire fabric. Plastic fibers are used for reinforcement only occasionally. With higher support demands of the ground and with shotcrete thicknesses of generally 6 inches or greater lattice girders are embedded within the shotcrete. Occasionally and if needed by special support needs rolled steel sets are used in lieu of, or in combination with lattice girders. Initial support also includes all measures of rock reinforcement in rock tunnelling. A prototypical ESC cross section and longitudinal section respectively. A cross section without a closed invert on the left side and ring closure on its right side.

Invert closure is typically required in weak rock conditions and squeezing ground. Figure includes elements of typical initial support including rock bolts/dowels, initial shotcrete lining and tunnel pre-support. The arrangement of rock bolts/dowels is typical and varies depending on the excavation and support.

Longitudinal Tunnel Profile and Distribution of Excavation and Support Classes (ESCs)

SEM contract documents contain all Excavation and Support Classes (ESCs) assigned along the tunnel alignment in accordance with the Ground Response Classes (GRCs) and serve as a basis to estimate quantities. A summary longitudinal section along the tunnel alignment shows the anticipated geological conditions, the GRCs with the relevant description of the anticipated ground response, hydrological conditions and the distribution of the ESCs.

Geological data, Ground Response Classes, Excavation and Support Classes, the Longitudinal Tunnel Profile as well as design assumptions and methods shall be described and displayed in reports that become part of the contract documents. When defining the reaches and respective lengths of GRCs and corresponding ESCs it is understood that these are a prognosis and may be different in the field. Therefore contract documents establish the reaches as a basis and call for observation of the ground response in the field and the need for their adjustment as required by actual conditions encountered. Actual conditions must be accurately mapped in the field to allow for a comparison with the baseline assumptions portrayed in the GRCs. For that purpose standard form sheets are developed as portrayed for a typical SEM rock tunnel mapping.

Excavation Methods

During the history of application of the SEM/NATM, tunnelling methods for a wide variety of ground conditions have been developed. With the further development and refinement of support means, the application field of the SEM has ever been expanded. From its original implementation in alpine, "green field" rock and soft rock tunnels the focus moved into urban areas and soft ground tunnelling. SEM tunnelling is typically accomplished in hard rock using drill-and-blast excavation techniques, medium hard and soft rock using a road header (6.4.3) and in soft ground using backhoe excavation.

Such SEM excavations from hard rock through soft ground. Drilling of a face in a rock tunnel for a drill-and-blast excavation. A close up of the drilling at the face is shown, that also displays the shotcrete initial lining installed close to the face. The rock face has been sealed by a layer of flashcrete. A close up of a road header boom excavating a medium hard, jointed rock mass. Tunnel construction of a soft ground tunnel in a top-heading, bench, and invert excavation using backhoes. The backhoe is in the background at the tunnel face.

Sequential Excavation Method (SEM)

Ground Support Elements

This section addresses special ground support and material considerations that have evolved with the application of shotcrete supported SEM excavations.

Shotcrete

The original name for shotcrete was “Gunite” when it was used for the purpose of taxidermy by spraying mortar on wire frames in the US in the early 1900’s. In its early applications, sprayed dry mix material has also been used for the improvement of the fire resistance of timber supports in mines. During the course of the early 1930’s, the term “Shotcrete” was introduced and has been widely used since. Development of equipment technology for the application of shotcrete progressed rapidly and the use of shotcrete for ground support purposes spread worldwide. In particular, the use of NATM / SEM, and the associated extensive use of shotcrete contributed to development of shotcrete which nowadays can be viewed as sprayed concrete, the major distinction between concrete and shotcrete being merely the method of placement (Vandewalle, 2005).

Effect of Shotcrete

When concrete is sprayed on a rough ground surface, it fills small openings, cracks and fissures and as initial support provides immediate support after excavation. It reduces the potential for relative movement of rock bodies or soil particles and, therefore, limits loosening of the exposed ground surrounding the tunnel. The adhesion depends on the condition of the ground surface, the dampness and presence of water and the composition of the shotcrete. Generally, the rougher the ground surface the better the adhesion. Dry rock surfaces have to be sufficiently dampened prior to application of shotcrete. Dusty or flaky surfaces, water inflow or a water film on the rock surface or other contaminant reduce the adhesion of shotcrete.

Modern admixtures improve the “stickiness” of shotcrete significantly such that rebound is reduced considerably. Fibers increase the adhesion and cohesion of the freshly applied (“green”) shotcrete and therefore improve the build-up quality of the shotcrete. In turn, excessive stickiness of the shotcrete mix (as frequently observed when sodium silicate accelerators are used) can have an adverse effect. Too sticky shotcrete tends to accumulate around reinforcement bars, resulting in insufficiently compacted, low quality concrete or even voids or “shadows” behind the reinforcement bars.

In order to stabilise small wedges and slabs, shotcrete is applied locally. This application type does not form a continuous layer of shotcrete over an extended area to form a supporting member in the sense of a lining or structural shell. Rather, edges and corners generated by the intersection of discontinuities are filled with shotcrete bonding the bodies together thus forming local support.

Shotcrete Face Support: In poor ground conditions a temporary face support may be required to restrict the ground from moving into the excavation. Dependent on the length of period through which the support is required and the ground conditions, the thickness and reinforcement of the face support varies. For tunnel stubs, permanent head walls are constructed with shotcrete. A domed face shape is of great importance in poor ground for successful face stabilisation.

Experience gained from tunnel projects in soft ground demonstrates that ground deflections and hence surface settlements continue until a final, fully domed head wall with sufficient connection to the tunnel shotcrete initial lining is established.

Temporary Shotcrete Support: In poor ground conditions or where large tunnel cross sections are constructed, the excavation area must often be split into several drifts. To provide immediate support and, if required, ring closure for each sub-drift, temporary shotcrete support shells or linings are used. The thickness of the temporary lining is designed based on the cross sectional area of the drift to be supported and the period for which the support is required. The temporary shell is removed during subsequent construction steps that complete the excavation to the full tunnel opening.

Figure: Typical Tunnel Excavation with Temporary Middle Wall (Beacon Hill Station, Washington)

Initial Shotcrete Lining: Initial shotcrete lining typically consists of 4 to 16 inch (100 to 400 mm) thick shotcrete layer mainly depending on the ground conditions and size of the tunnel opening, and provides support pressure to the ground. It is also referred to as shotcrete lining. A shotcrete ring can carry significant ground loads although the shotcrete lining forms a rather flexible support system. This is the case where the shotcrete lining is expected to undergo high deformations and hence ductility post cracking is of importance. By deforming, it enables the inherent strength and self-supporting properties of the ground to be mobilised as well to share and re-distribute stresses between the lining and ground. During deformation, stresses acting within the shotcrete lining are transferred into the surrounding ground. This process generates subgrade reaction of the ground that provides support for the lining. From the ground support point of view the design of the shotcrete lining is governed by the support requirements, i.e., the amount of ground deformations allowed and ground loads expected as well as economical aspects. The earlier the sprayed concrete gains strength the more the support restrains ground deformation. However, by increasing stiffness the support system increasingly attracts loads. It depends on the ground conditions and local requirements how stiff or flexible the support system has to be and thus what early strength requirements, thickness and reinforcement should be specified.

In shallow tunnel applications and beneath surface structures that are sensitive to deformations such as buildings, ground deformations and consequently surface settlements have to be kept within acceptable limits. The advantage of the mobilisation of the self-supporting capacity of the ground can therefore be only taken into account to a very limited extent. Here, early strength of the shotcrete is required to gain early stiffness of the support to limit ground deflections. Under these conditions the shotcrete lining takes on significant ground loads at an early stage however in a generally low stress environment due to the shallow overburden. Early strength can be achieved with admixtures and modern cement types.

In contrast for tunnels under high overburden the prevention of ground deformation and surface settlement plays a secondary role. Excessive ground loads in squeezing ground and active tectonic pressures applied on the tunnel perimetre may be the design criteria for deep tunnels. By allowing the ground to deflect (without over-straining it) the ground's self-supporting capability, mainly shear strength, is mobilised. Consequently, the ground loads acting upon the shotcrete lining can be limited significantly because the ground assumes a part of the support function and a portion of the ground loads is dissipated before the initial support is loaded. For rock tunnels under high cover, early strength is not a necessity but final strength of the entire system is of importance. In special cases it may be required to construct "deformation joints" by implementing special yield elements to allow substantial deformation of the ground without generating uncontrolled cracks in the shotcrete lining while maintaining a defined support pressure during the process of deformation. Yield elements are designed such as to allow prescribed maximum deformation under defined lining loads.

Rock reinforcement installed in rock tunnels augments the strength of the surrounding ground, controls deformation and limits the ground loads acting upon the shotcrete initial lining. Shotcrete support and rock reinforcement are designed to form an integrated support system in view of the excavation and support sequence. The design engineer must define the requirements for the support system based on thorough review of the ground response anticipated.

The effect of the shotcrete is heavily dependent on the radial and tangential subgrade reaction generated by the surrounding ground. Therefore, shape, shotcrete thickness and installation time have to be designed in accordance with the ground conditions and the capacities of the surrounding ground and the support system. Site personnel should assess the support requirements and, if necessary, adjust the designed support system based on observations in the field. Notwithstanding the need for reaction to site conditions, the designer should always be party to the decision making process prior to changing any support means on site. The design intent and philosophy must be taken into consideration when adjustments of the support system are made.

Friction between the ground and the sprayed concrete lining (tangential subgrade reaction) is paramount for the support system. This friction reduces differential movement of ground particles at the ground surface and contributes to the ground-structure interaction. Even the shotcrete arch not forming a closed ring provides substantial support to the ground, given that tight contact between the sprayed concrete and the ground is maintained.

The requirement for a ring closure, be it temporary or permanent, is governed by the size of the underground opening and the prevailing ground conditions. In a good quality rock mass, no ring closure is required. In low quality ground (weak rock and soil), it has been proven in numerous case histories that the time of support application after excavation, length of excavation round and time lag between the excavation of the top heading and the invert closure rules the ground and lining deflection. To reduce ground deflection and the potential for ground/lining failure, the excavation and support sequence must be designed such that an early ring closure of the shotcrete support in soft ground is achieved. Also the timely (immediately after excavation) installation of the shotcrete support members is of utmost importance. To achieve an early, temporary ring closure and to reduce excavation face size, partial drifts such as sidewall drifts, middle drifts and top heading, bench, and invert drifts can be used. These partial drifts are supported by temporary shotcrete support, such as temporary middle walls, invert supports, etc.

An important aspect of shotcrete linings is the design and construction of construction joints. These joints are located at the contact between shotcrete applications in longitudinal and circumferential directions between the initial lining shells of the individual excavation rounds and drifts. An appropriate location and shape as well as connection of the reinforcement through the longitudinal joints is of utmost importance to the integrity and capacity of the support system. Longitudinal joints have to be oriented radially, whereas circumferential joints should be kept as rough as possible. Splice bars/clips and sufficient lapping of reinforcement welded wire fabric maintain the continuity of the reinforcement across the joints. Rebound, excess water, dust or other foreign material must be removed from any shotcrete surface against which fresh concrete will be sprayed. The number of construction joints should be kept to a minimum.

In case of ground water ingress, the ground water has to be collected and drained away. Any build-up of groundwater pressure behind the shotcrete lining should be avoided for the following reasons: Increased ground water pressure in joints and pores reduces the shear strength in the ground, undue loads may be shed onto the shotcrete lining (unless it is designed for that, which is unusual for initial shotcrete linings); softening of the ground behind the lining; increased leaching of shotcrete; shotcrete shell will be detached from the ground.

Rock Reinforcement

Rock reinforcement and rock mass act as a complex interactive system, where the individual elements always have to be seen in view of their interaction and interdependence. The overall strength of a reinforced rock mass with a joint system is governed by the characteristics of the joints (roughness, fill, rock material, orientation) and the contribution provided by the reinforcement elements. For the design of rock mass reinforcement systems, sufficient appreciation of the expected ground conditions and experience are of fundamental importance. The following focuses on the SEM applications and issues.

Types of Rock Reinforcement

Rock Dowels: Passive reinforcement elements that require some ground displacement to be activated. In deep tunnels or under tunnelling conditions where ground deflection is permitted or even desired, passive rock reinforcement is frequently installed. This applies for example to tunnel construction sequences where the excavation and support installation is carried out in sequences (e.g. top heading, bench, invert). In order to best use the support effect of the rock dowels, an early installation is required. The majority of ground deflections develop during excavation and closely behind the progressing tunnel face. In sequential rock tunnelling using multiple drifts, ground deflection typically ceases after top heading excavation and support but commences again after a period of relative stability during excavation for bench and invert construction. Therefore, rock dowels should be installed right after excavation or close to the progressing excavation face.

Rock Bolts: Actively introduce a compressive force into the surrounding ground. This axial force acts upon the rock mass discontinuities thus increasing their shear capacity and is generated by pre-tensioning of the bolt. The system requires a 'bond length' to enable the bolt to be tensioned. Rock bolts frequently are fully bonded to the surrounding ground after tensioning, for long-term load transfer considerations.

Rock bolts are not only installed during construction. Rock bolts may also be used for existing under-ground openings, where further deformation of the ground and/or the support is to be inhibited or for additional support of existing structures that will undergo subsequent enlargement or be influenced by adjacent tunnel construction.

For tunnels constructed in an environment where ground deflection and surface settlement hast to be limited (e.g. shallow tunnels in urban areas), rock bolt aid in limiting the ground displacement caused by the SEM tunnelling. Furthermore, during construction of large openings ground deflection limitation may be desired to avoid loosening (and hence weakening) of the rock mass. In high stress environments, special compressible elements have been developed, that are installed between the ground/support surface and the face plate of the bolt allowing a certain amount of displacement while the tension force at the bolt is kept constant.

Rock Anchors: Rock anchors are used under conditions where high anchor forces have to be accommodated often significantly higher than for example rock bolt forces. For instance in very large span tunnels, where high support forces have to be generated to stabilise the ground, anchors are frequently used.

Generally, it can be stated that pre-tensioning of bolts establishes a stiffer system of the reinforced rock mass after installation and minimises the magnitude of shear displacement. The design and application of a pre-tensioned rock reinforcement system requires excellent knowledge of the ground conditions and ground behaviour to avoid over-tensioning during ground displacement. In comparison, an initially untensioned rock dowel reinforcement may ultimately lead to the same strength and reinforced rock mass capacity, however, only along with larger deformations.

Lattice Girders and Rolled Steel Sets

Lattice girders are lightweight, three-dimensional steel frames typically fabricated of three primary bars connected by stiffening elements. Lattice girders are used in conjunction with shotcrete and once installed locally act as shotcrete lining reinforcement. The girder design is defined in the contract documents by specifying the girder section and size and moment properties of the primary bars. To address stiffness of the overall girder arrangement the stiffening elements must provide a minimum of five percent of the total moments of inertia. This percentage is calculated as an average value along repeatable lengths of the lattice girder. The arrangement of primary bars and stiffening elements is such as to facilitate shotcrete penetration into and behind the girder, thereby minimising shadows. Lattice girders are installed to provide:

- Immediate support of the ground (in a limited manner due to the low girder capacity)
- Control of tunnel geometry (template function)
- Support of welded wire fabric (as applicable)
- Support for fore poling pre-support measures

In particular cases where, for example, immediate support is necessary for placing heavy spilling for pre-support, the use of rolled steel sets may be appropriate. In such instances steel sets are used for implementation of contingency measures. Steel sets of bell shaped profile (Heintzmann profile) are also used as structural members in temporary shotcrete sidewalls in multiple drift tunnelling. Their primary purpose apart from increased capacity over lattice girders is their ease of removal when demolishing temporary shotcrete walls in multiple drift tunnelling applications.

Pre-support Measures and Ground Improvement

When tunnelling in competent ground, the ground surrounding the tunnel opening provides sufficient strength to ensure stand-up time needed for the installation of the initial SEM support elements without any pre-support or improvement of ground strength prior to tunnelling. With the significantly increased use of the SEM in particular in soft ground and urban areas over the past decades, traditional measures to increase stand-up time were adopted and further developed to cope with poor ground conditions and to allow an efficient initial support installation and safe excavation.

These measures are installed ahead of the tunnel face. They include ground modification measures to improve the strength characteristics of the ground matrix including various forms of grouting, soil mixing and ground freezing, the latter for more adverse conditions. Most commonly they include mechanical pre-support measures consisting of spiling methods installed ahead of the tunnel face often with distances of up to 60 to 100 feet (18 to 30 m) referred to as pipe arch canopies or at shorter distances, as short as 12 ft (3.6 m) utilising traditional spiling measures such as grouted solid bars or grouted, perforated steel pipes. Ground improvement and pre-support measures can be used in a systematic manner over long tunnel stretches or only locally as required by ground conditions.

Pre-support Measures

Pre-support measures involve spiling or grouted pipe arch canopies that bridge over the unsupported excavation round. These longitudinal ground reinforcement elements are supported by the previously installed initial shotcrete lining behind the active tunnel face and the unexcavated ground ahead of the face. These mechanical pre-support measures are generally used to:

- Increase stand-up time by preventing ground material from raveling into the tunnel opening causing potentially major over-break or tunnel instabilities

- Limit over-break
- Reduce the ground loads acting on the immediate tunnel face
- Reduce ground deflection and, consequently surface settlements.

Mechanical pre-support measures are generally less intrusive than systematic ground modifications. They rely on the ground reinforcing action of passive reinforcement elements such as steel or fiberglass pipes/bars. Similar to passive concrete reinforcement the elements must directly interact with the surrounding ground to be efficient as reinforcement. This interaction can only be established by a tight contact between the reinforcement element and the ground. This interaction can be achieved by either fully grouting the pre-support elements to lock the reinforcement in with the ground or by ramming the reinforcement elements into ground if susceptible to this action in soft ground conditions. Loosely installed elements installed in soft rock or soil do not achieve their intended function and such installations must be avoided. In fractured, but competent rock, steel rebars loosely installed in boreholes may be acceptable but merely to limit over-break. Spiles rest on the initial shotcrete lining (front) and on the unexcavated ground beyond the tunnel face. The narrow spacing allows even very soft and soils with little cohesion to bridge between individual spiles.

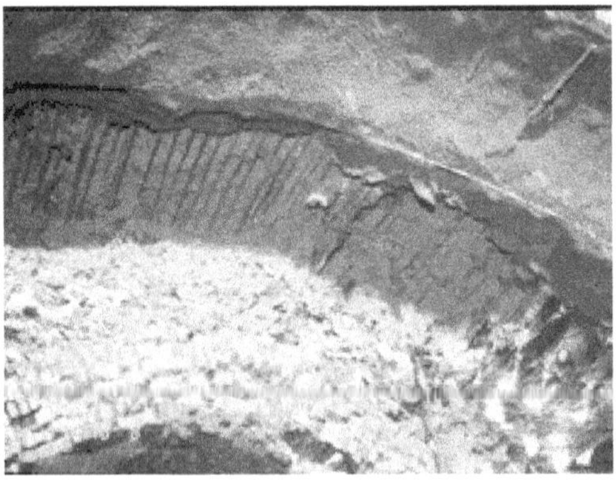

Figure: Spiling Pre-support by No. 8 Solid Rebars (Berry Street Tunnel, Pennsylvania)

The effect of mechanical pre-support has frequently been misjudged. On one hand, the stiffness of the steel elements used for pre-support is often taken as basis for assessing an increase of the overall stiffness of the ground surrounding the pre-support. This can easily lead to an over-estimation of the pre-support function, as the longitudinal stiffness of the entire system must be taken into account in those considerations. On the other hand, the radial action of a systematic pre-support arch is often underestimated or not considered at all. The longitudinal effect of a pre-support element is less governed by the stiffness of the reinforcement element than by the improvement of the tensile and shear capacity of its surrounding ground.

When grout is used to establish the bond between the reinforcing element and the ground, grouting pressure used for installation, type of grout and grouted length have a paramount influence on the effect and efficiency of the pre-support in particular in soft ground conditions. Though it has been proven in countless applications that mechanical pre-support has the effects mentioned above, quantification of the effect by numerical analyses methods proved to be difficult involving efforts that go beyond the usual design efforts. Hence, the effect of pre-support is often assessed using simple approaches that result in very conservative assessments, thus underestimating the actual effect of pre-support. In many cases, the effect of pre-support is even ignored in a design and pre-support is viewed merely as an increase of the safety margin rather than a settlement-limiting element of the tunnel support.

Pre-Support in Rock Tunnelling: Pre-support installation in fractured, yet competent rock mass types is typically aimed at limiting the over-break during and after excavation. Pre-spiling with steel rebars is a frequent method to keep rock fragments in place. Dependent on the degree of fracturing, the rebars are installed in empty boreholes arranged around the perimetre of the roof, or the boreholes are filled with cement grout prior to insertion of the rebars. Alternatively, perforated steel pipes are used that are inserted into the boreholes and subsequently grouted. In a severely fractured rock mass where boreholes tend to collapse, self-drilling rock reinforcement pipes are used. With the grouted applications, grout may intrude into cracks and fractures introducing a limited cementing effect of the surrounding material. In soft rock mass types, where fracturing and limited material strength result in conditions with low overall strength, grouted pipe spiling or grouted pipe arches are used for pre-support. If required, these pre-support measures are combined with groundwater drawdown measures to reduce the joint water pressure and to increase the frictional capacity along the joints.

Permeation grouting of the discontinuities is used to reduce the mass permeability and to increase the overall shear strength by cementing the rock fragments together.

Pre-Support in Soft Ground (Soil) Tunnelling: Similar to soft rock, grouted mechanical pre-support measures are used to pre-stabilise soil or soil like ground. Dependent on the susceptibility of the soil to grout, these mechanical pre-support methods are combined with grouting systems that allow penetration of grout into the ground leading to cementation of the ground surrounding the pre-support. Penetrability of the ground and the intended purpose of the pre-support govern the selection of the grouting materials. While grout with standard cements has a limited capability for penetrating ground containing sand or smaller fractions, penetration results can be improved by the use of micro or ultra fine cement products or chemical grouting (resin grouting). The current market offers resin grouting materials with viscosity values close to water.

In many cases, particularly under shallow cover with the groundwater table in the lower part or below the tunnel invert, mechanical pre-support measures are sufficient as long as the support elements are sufficiently locked into the ground over their entire length by an appropriate grouting material. Any additional effect by grout material penetrating voids in vicinity of the installed pre-support is considered an additional benefit. In very loose, generally non-cohesive ground, ground improvement measures may be required to cement the ground and to decrease the permeability of the soil.

Pre-Support Elements: Most commonly used mechanical pre-support elements include grouted pipe spiling of typically 2-inch (50 mm) diametre perforated steel pipes and rebar spiling using solid No. 8 (25 mm diametre) steel rebars as shown. These are primarily installed in the area of the tunnel roof and shoulders, but may also be installed in the sidewall and invert if suitable and required. Grouting of these spiling elements establishes a tight contact between the reinforcement element and the surrounding ground. So-called self-drilling and grouted rebars (type IBO, ISCHEBECK or similar) provide for a very efficient installation of grouted, solid steel bars.

Grouted Pipe Arch Canopy: Pipe arch canopy methods involve a systematic installation of grouted pipes at a spacing of typically 12 inch (300 mm) around the tunnel crown. This installation typically involves one single row of pipes but under critical ground conditions and / or when surface settlements must be restricted may involve a double row of pipes. The pipes are installed at lengths typically not to exceed 15 to 24 metres (50 to 80 feet) using conventional drilling techniques at a shallow lookout angle from the tunnel and ahead of the tunnel excavation. Specialised drill bit and casing systems are utilised that aim at limiting and strictly controlling the over cut, i.e., annular void space between inserted pipe and the surrounding ground. They also provide for direction control and high installation accuracy. Drilling techniques include ODEX®, CENTREX®, ALWAG and similar methods.

The steel pipes are typically perforated and have a diametre of between 4.5 inch and 6 inch (114 mm to 150 mm). The steel pipes are grouted to facilitate contact between steel pipe and the surrounding ground and to create the desired arching effect around the tunnel opening during excavation. Depending on purpose and susceptibility of the ground to grouting, the perforated steel pipes may be grouted either from the single entry point at pipe end within the tunnel or using packers or double packers. Grouting with double packers will allow for targeted grouting with respect to location, grout mix, injected volumes, and pressures. These pipe arch systems have furthered the use of SEM applications in particular in urban settings under shallow overburdens and also in difficult ground conditions.

The installation of a steel pipe for an arch application for a 3-lane road tunnel in soft ground. The steel pipe on a drill jumbo boom and a 4.5 inch (114 mm) steel pipe being drilled near the circumference of the shotcrete initial lining. Previously installed pipe arch steel pipes exposed in the ground when opening a new excavation round.

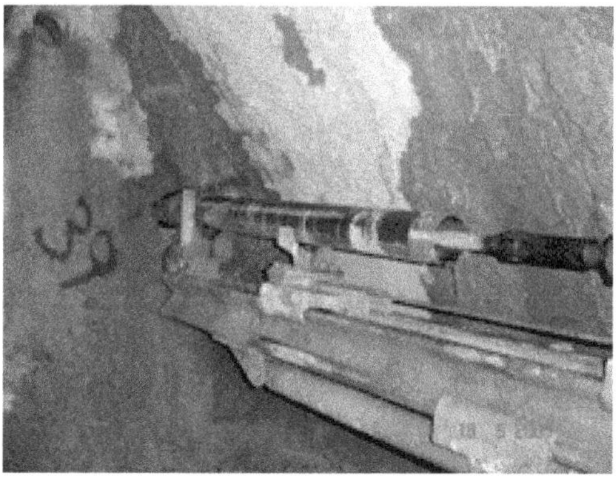

Figure: Steel Pipe Installation for Pipe Arch Canopy

Face Doweling: Face doweling forms a specific form of pre-support. Other than the mechanical pre-support installed in the tunnel roof and shoulder area, the face pre-support is installed within the excavation face to stabilise squeezing or raveling ground at the face prior to excavation. Passive elements are installed in the ground and usually grouted in place to increase the tensile and shear strength of the ground material. Since the reinforcement elements have to be excavated during subsequent excavation rounds, fiberglass reinforced resin dowels or pipes are frequently used. Steel elements for face doweling hamper the excavation progress and during excavation their removal transfers tension forces into the ground, promoting ground disturbance ahead of the progressing tunnel face. Face doweling can be combined with application of grouting methods to locally improve the overall strength of the ground within the tunnel cross-section and act with the face dowels.

Face support dowels are usually made of GFRP (glass fibre reinforced polyester resin) and provide significant tensile strength while allowing for easy removal during excavation due to the material composition and low shear resistance.

Ground Improvement

Ground improvement measures are primarily aimed at modifying the ground matrix to increase its shear (cohesion) and compressive strengths. An increase of the stiffness (deformation modulus) is coincidental to this improvement. These measures are frequently installed from the surface and well in advance of the tunnel excavation or are applied from within the tunnel ahead of the face. Ground improvement measures range from lowering of the groundwater table or reduction of the pore/joint water pressure to intrusive changes of the ground composition such as jet grouting, soil mixing or ground freezing.

Groundwater Draw Down: Draw down of the groundwater table reduces or eliminates the groundwater inflow into tunnels during construction and increases the effective shear strength of the ground. Groundwater flowing into the tunnel opening during construction not only causes unsafe conditions and increases equipment wear and tear; it also can promote ground instabilities. The reduction of the hydrostatic head reduces the water pressure acting within discontinuities and soil pores. Groundwater draw down can be carried out from the surface or from within the tunnel.

In fine-grained soils (fine sands, silts, clays) the reduction of the pore pressure results in a significant increase of the overall strength of the ground. Where gravity drainage is insufficient, vacuum wells or other means such as drainage by osmosis can be applied.

Permeation Grouting: Permeation grouting is frequently used to cement the ground matrix if it is sufficiently coarse and uniform to achieve reliable grout penetration. Microfine cement or chemical (resin) grouts are used for finer grained soils.

Where soils are not sufficiently uniform or groutable, other measures such as jet grouting or soil mixing are used. These methods actively modify the ground's fabric by mixing the ground with a cementing agent such as cement grout or lime. Jet grouting uses a high-pressure water-grout mix jet to cut the ground and mix it with the stabilising agent generating improved soil columns of significant diametre.

Ground Freezing: Ground freezing is often considered as 'last resource' due to its high cost when compared to other ground improvement measures. However, ground freezing achieves a high degree of reliability of ground modification. This particularly applies for non-uniform soils. The frozen ground provides groundwater cut-off while its mechanical properties are sufficiently increased to allow an efficient and safe tunnel excavation and support installation under the protection of the frozen soil body. Ground freezing has provided solutions for tunnelling under very complex conditions in urban settings.

Portals

This section describes the layout of temporary tunnel portal structures for highway tunnels that are frequently built with SEM tunnelling. These structures provide a protection against rock fall, and stabilise the portal face from which SEM tunnelling commences thus provide start-up condition for safe tunnel excavation.

Shotcrete canopies are also frequently used as an extension of the tunnel and are integrated into the final tunnel portal architecture. The tunnel final lining is cast against these shotcrete canopies and therefore the tunnel internal geometry is uniform from the cut-and-cover (shotcrete canopy) section into the mined tunnel. The shotcrete canopies are backfilled for the final condition.

Pre-Support and Portal Collar

The level of weathering and loosening of rock close to the surface must be addressed when starting tunnel construction. Even in generally good rock mass, surface near weathering and loosening requires pre-support at the portal.

After clearing the surface and installing required rock support at the portal face, a row of horizontal pre-spiling or grouted steel pipes should be installed to provide pre-support for the initial excavation rounds for the tunnel construction. Dependent on the degree and depth of weathering, this pre-support is typically 10 ft (3 m) to 60 ft (18 m) long and the reinforcement elements are grouted in place. The pre-support elements are typically spaced at 12-inch (0.30 m) centres around the future tunnel opening.

Figure: Pre-Support at Portal Wall and Application of Shotcrete for Portal Face Protection (Devil's Slide Tunnels, California)

Following the pre-support installation, a reinforced shotcrete collar should be installed that is tied in with the protruding pre-support elements. The collar shall follow the tunnel perimetre extending from one sidewall to the other. In soft ground, the collar may extend over the entire tunnel perimetre. The collar provides stability to the ground in the immediate vicinity of the future tunnel opening and is structurally connected to the initial shotcrete lining for the first round of tunnel excavation.

Shotcrete Canopy

The shotcrete canopy comprises reinforced shotcrete and lattice girders. The canopy is founded on a strip foundation that extends over the entire length of the canopy. The length of the canopy is dependent on the rock fall protection required and on local conditions such as wind loads, temporary ventilation requirements, and needs of the final tunnel structure.

Portal canopies have to be designed for rock fall and snow loads, construction loads, dead loads, and any wind loads, as dictated by local site conditions. The canopy also serves as a counter form for final lining installation in the portal area. Construction of a shotcrete canopy whereas the first three lattice girders and reinforcement have been placed and shotcrete is being sprayed against an expanded metal sheet placed on the outside of the lattice girders.

Figure: Shotcrete Canopy Construction after Completion of Portal Collar and Pre-support

Structural Design Issues

Ground-Structure Interaction

The SEM realises excavation and support in distinct stages with limitations imposed on size of excavation and length of round followed by the application of initial support measures. In particular the shotcrete lining has an interlocking function and provides an early, smooth support. To adequately address this sequenced excavation and support approach the structural design shall be based on the use of numerical, i.e. finite element, finite difference, or distinct element methods. These numerical methods are capable of accounting for ground structure interaction. They allow for representation of the ground, the structural elements used for initial and final ground support, and enable an approximation of the construction sequence.

Embedded frame analyses have limitations in adequately describing the ground structure interaction. Due to this and the fact that these methods can not simulate excavation sequencing their use shall be limited to applications where the ground structure interaction phenomenon, in particular development of a ground-supporting arch, is of secondary importance. This is for example the case for shotcrete canopies that are frequently erected at tunnel portals as freestanding or backfilled reinforced shell structures and tunnel final linings.

Numerical Modelling

Two (2)-Dimensional and Three (3)-Dimensional Calculations

In general, use of two-dimensional models is sufficient for line structures. Where three-dimensional stress regimes are expected, such as at intersections between main tunnel and cross passages, or where detailed investigations at the tunnel face are undertaken such as for the behaviour of pipe arch pre-supports, three-dimensional models should be used.

Material Models

In representing the ground, the structural models shall account for the characteristics of the tunnelling medium. Material models used to describe the behaviour of the ground shall apply suitable constitutive laws to account for the elastic, as well as inelastic ranges of the respective materials. For example when tunnelling in rock, intact rock as well as the rock structure, i.e., the presence of discontinuities shall be taken into account. It is customary to apply Mohr-Coulomb or Drucker-Prager failure criteria for the representation of both rock and soil materials. Finite element programs that were developed initially for the simulation of underground excavations in rock such as Phase 2 by Rockscience, Inc. also allow use of rock mass material behaviour using Hoek and Brown rock mass parameters (Hoek and Brown, 1980, 2002).

Ground Loads - Representation of the SEM Construction Sequence

Tunnel excavation causes a disturbance of the initial state of stress in the ground and creates a three-dimensional stress regime in the form of a bulb around the advancing tunnel face. Far ahead of the advancing tunnel face the initial state of stress is represented by vertical and horizontal stress trajectories denoting major and minor principal stresses respectively (assuming that vertical stresses are higher than horizontal stresses in a geostatic stress field). At the tunnel face the stresses flow around the tunnel opening arching ahead of the tunnel excavation and behind it onto the newly constructed initial lining in longitudinal direction and to the sides of the opening perpendicular to the tunnelling direction. At a distance where the tunnel is no longer affected by the three-dimensional stress conditions around the active tunnel face two-dimensional arching conditions are established.

The extent of the stress disturbance around an active heading depends mainly on ground conditions, size of the excavation and length of round. This disturbance begins up to two excavation diametres ahead of the active tunnel face as shown indicatively. The SEM design dictates limits on excavation size and length of round and prescribes installation of initial support elements often following each individual excavation round directly or shortly thereafter. These requirements are portrayed in the Excavation and Support Class. Initial support elements are therefore installed within the shelter of a load-carrying arch around the newly created opening in an area where some pre-deformation has occurred. As the excavation of the tunnel advances the shotcrete hardens from an initially "green" shotcrete and becomes fully loaded at a distance of about one to two tunnel diametres from the face. Such sequencing combined with early support installation contributes to the development of the self-supporting capability of the ground.

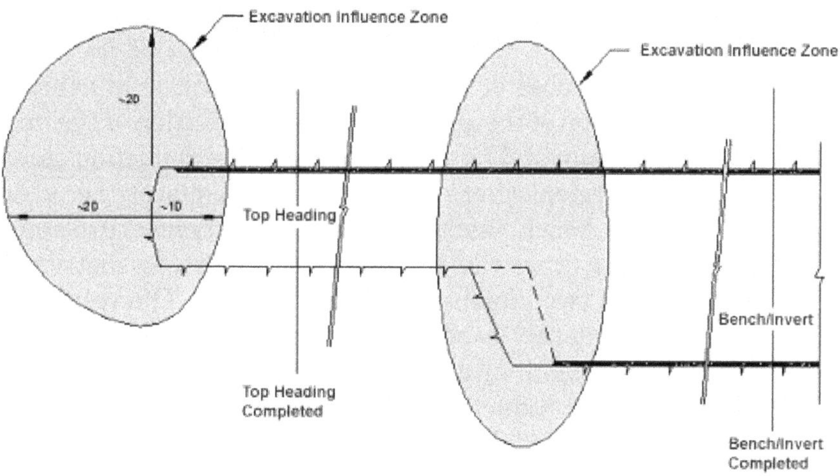

Figure: SEM Tunnelling and Ground Disturbance (after OGG, 2007)

It is therefore important to portray this excavation and support sequencing closely in the numerical analyses. For shotcrete lining structural assessments it is important to distinguish between a "green" shotcrete when installed and when it has hardened to its 28-day design strength. Green shotcrete is typically simulated using a lower modulus of elasticity in the computations. A value of approximately 1/3 of the elastic modulus of cured shotcrete is commonly used to approximate green shotcrete in 2-D applications. In 3-D simulations the shotcrete may be modelled with moduli of elasticity in accordance with the anticipated strength gain in the respective round where it is installed.

Excavation and support installation sequencing can be readily realised in three-dimensional models. In two-dimensional modelling, however, auxiliary techniques must be utilised. A frequently utilised approach relies on the use of ground modulus reduction within the excavation perimetre prior to the insertion of the initial lining elements into the model. Other techniques involve the use of supporting forces applied to the circumference of the tunnel opening. Because of its frequent use the ground modulus reduction approach is used in describing a typical two-dimensional modelling sequence of SEM tunnel excavation and support of a line structure below. A calculation example is provided.

- Represent the in-situ stresses including the geostatic stress field and surface loads as applicable.
- Represent the excavation of the respective round by reducing the elastic modulus of the ground located within the geometric boundaries of the round to about 40% - 60% of its original value. The purpose of the modulus reduction is to achieve a pre-deformation of the ground prior to installation of the initial support measures. The extent of modulus reduction is only within the region where excavation takes places, i.e. a drift (top heading, bench, invert). It is an arbitrary measure applied to simulate an otherwise three-dimensional stress distribution at the face in two-dimensional computations. The value of 40-60% is a frequently used reduction amount and represents a typical range (Mohr and Pierau, 2004 and Coulter and Martin, 2004). A higher reduction will yield larger, a lower reduction will yield smaller deformations of the surrounding ground. A sensitivity analysis related to the actual reduction value is typically part of the computations.
- Activate the initial support elements per design assumptions to represent the installation of initial support. Because the shotcrete will not have developed its design strength at this stage reduced shotcrete elastic properties (modulus) are initially taken into account and amount to about 1/3 of the hardened shotcrete. During subsequent simulation stages the shotcrete modulus is then increased to its 28-day design strength to represent a fully hardened shotcrete lining.
- Remove the ground elements within the respective drift thereby completing excavation and support within the round.
- Repeat this sequence until all drifts of the final tunnel cross section geometry have been excavated and supported.

Once accomplished, this completes modelling of the tunnel excavation and installation of initial support. The installation of the final tunnel lining generally occurs once all deformations of the tunnel opening have ceased. To account for this fact the calculations perform installation of the final lining into a stress-free state. The final lining becomes loaded only in the long-term resulting from a (partial) deterioration of the initial support (shotcrete initial lining and rock bolts if any), rheological long-term effects and ground water if applicable. Although modelling of the final lining is often undertaken by embedded frame analyses its analysis within a ground-structure interaction numerical model will be most appropriate and can follow directly after the initial support is installed as follows:

- Activate the structural elements representing the final tunnel lining.
- If the modelling was carried out with temporary rock reinforcing elements without corrosion protection then all such supporting elements are deactivated.
- If the ground water is generally aggressive and it may be assumed that the shotcrete initial lining will deteriorate long-term then it is assumed that no contributing support function may be derived from it for long-term considerations. This has been traditionally assumed on projects such as the Lehigh Tunnel, Cumberland Gap Tunnels and on NATM tunnels of the Washington, DC Metro. Washington Metropolitan Area Transit Authority (WMATA) has substantial experience with the design and construction of NATM tunnels in both soft ground and rock (Rudolf et al., 2007). To date it is customary on WMTA projects to assume that the shotcrete initial lining will deteriorate over time. Such computational approach will yield a conservative final lining design. Due to the nowadays high quality shotcrete fabrication however, and in particular in non-aggressive ground and ground water conditions it is admissible to assume that when the shotcrete initial lining is more than approximately 6-inch (150 mm) thick then 50% of its structural capacity may be taken into account in the final lining computations. The combined removal of initial support elements (rock reinforcement and shotcrete initial lining) will result in ground loads imposed onto the final lining in the long-term.
- In addition to the ground loads, the concrete lining will be loaded with hydrostatic loads in un-drained or partially drained waterproofing systems. This load case generally occurs well before the final lining is loaded with any ground loads and shall be considered separately in the calculations.

- Final lining calculations consider the existence of the waterproofing system, which is embedded between the initial shotcrete lining and the final lining. A plastic membrane will act as a de-bonding layer in terms of the transfer of shear stresses. Therefore simulation techniques should be used to simulate this "slip" layer. This is accomplished by only allowing the transfer of radial forces from the initial lining onto the final lining.

Ground Stresses and Deformations

Each step involving the simulation of excavation and installation of initial support allows for analysis of the ground response expressed in deformations, strains, and stresses. Both, elastic and, if yielded, inelastic portions of strains can be obtained and used to evaluate the state of stress in the ground and its capacity reserves. Stresses, strains and section forces are available in ground support elements such as dowels and rock bolts. The computational programs often provide such information in a user friendly display using numeric and graphic formats.

Lining Forces

Section forces and stresses are available for beam (2-D) or shell (3-D) elements. Section force and moment combinations are used to evaluate the capacity of the initial shotcrete and final concrete linings using ACI 318 or other accepted concrete design codes. Acceptance of codes is generally an owner driven process. For example, Washington Metropolitan Area Transit Authority (WMATA) allowed the use of the German Industry Standard DIN 1045 for the design of plain (unreinforced) cast-in-place concrete final linings. Based on this evaluation the adequacy of lining thickness and its reinforcement (if any) is assessed. If the selected dimensions are found not to be adequate then the model must be re-run with increased dimensions and/or reinforcement. The process is an iterative approach until the design codes are satisfied. These calculations do not distinguish between the type of lining application and therefore shotcrete and cast-in-place final linings are treated in the same manner within the program using the material properties and characteristics of concrete.

Ground Reinforcing Elements

Ground reinforcing elements are rock bolts and dowels. These are activated in the computations in accordance with the design of the SEM excavation and support installation. Once implemented and loaded during the simulation of excavation and support, section forces and stresses are available to evaluate their adequacy. Stresses and forces are compared with the capacity of the individual dowel or bolt.

Calculation Example

A calculation example (Appendix F) demonstrates the SEM tunnelling analysis and lining design of a typical two-lane highway tunnel using the finite element code Phase2 by Rocscience, Inc. The calculation is carried out in stages and follows the approach laid out in 9.6.2.3 above and evaluates ground reaction as indicated in 9.6.2.4 and evaluates support elements as described in 9.6.2.5 and 9.6.2.6.

Considerations for Future Loads

Mainly due to its flexibility and ability to minimise surface settlements often in combination with ground improvement methods the SEM is frequently utilised for the construction of roadway tunnels in urban settings. In particular under such circumstances it is important to consider any future loads that may be imposed onto the tunnel for which the final linings must be designed. Such loads include among others buildings, foundations, and miscellaneous underground structures fulfilling future infrastructure needs. These can be readily implemented in the computation approach presented above in the form of external or internal modelling loads.

Instrumentation and Monitoring

An integral part of SEM tunnelling is the verification by means of in-situ monitoring of design assumptions made regarding the interaction between the ground and initial support as a response to the excavation process.

For this purpose, a specific instrumentation and monitoring program is laid out in addition to general instrumentation programs connected with the overall tunnelling work, i.e. surface and subsurface instrumentation. The SEM tunnel instrumentation aims at a detailed and systematic measurement of deflection of the initial lining. While monitoring of deformation is the main focus of instrumentation historically stresses in the initial shotcrete lining and stresses between the shotcrete lining and the ground were monitored to capture the stress regime within the lining and between lining and ground. Reliability of stress cells, installation complexity and difficulty in obtaining accurate readings have nowadays led to the reliance on deformation monitoring only in standard tunnelling applications. Use of stress cells is typically reserved for applications where knowledge of the stress conditions is important, for example where high and unusual in-situ ground stresses exist or high surface loads are present in urban settings.

Monitoring data are collected, processed and interpreted to provide early evaluations of:

- Adequate selection of the type of initial support and the timing of support installation in conjunction with the prescribed excavation sequence
- Stabilisation of the surrounding ground by means of the self-supporting ground arch phenomenon,
- Performance of the work in excavation technique and support installation
- Safety measures for the workforce and the public
- Long-term stress/settlement behaviour for final safety assessment
- Assumed design parameters, such as strength properties of the ground and in-situ stresses used in the structural design computations.

Based on this information, immediate decisions can be made in the field concerning proper excavation sequences and initial support in the range of the given ground response classes (GRC) and with respect to the designed excavation and support classes (ESC).

Surface and Subsurface Instrumentation

The general instrumentation should include surface settlement markers, cased deep benchmarks, sub-surface shallow and deep settlement indicators, inclinometres, multiple point borehole extensometres, and piezometres.

The locations, types and number of these instruments should be determined by consultations between the civil, structural, geotechnical and SEM design teams to provide information on surface and subsurface structure settlements and to complement the SEM tunnel instrumentation readings.

Tunnel Instrumentation

Deformation Measurements Instruments are installed in the tunnel roof and at selected points along the tunnel walls to monitor vertical, horizontal, and longitudinal (in tunnel direction) deformation components. The number of points and their detailed location depends on the size of the tunnel and the excavation sequencing in multiple drift applications.

As a minimum, the wall of each drift (including temporary) should be equipped with a device capable of measuring deformations. It is customary to install optical targets for this purpose.

Figure shows a series of deformation monitoring cross sections using optical targets in a SEM tunnel. Optical targets are the white reflecting points arranged in the tunnel roof and tunnel sidewalls.

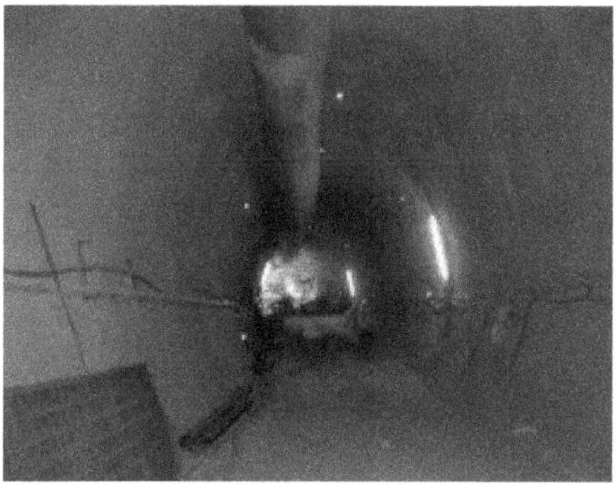

Figure: Deformation Monitoring Cross Section Points

Stress Measurements If stress information is sought then measurements should be taken with a direct measuring tool that does not rely on any further conversions from say strains to stresses. For example, instruments based on strain gage principles require the knowledge of the elastic modulus of the material to covert strains to stresses. This introduces an additional parameter that must be estimated thus introducing a secondary uncertainty.

Stress measurements within shotcrete linings are frequently carried out using hydraulic pressure cells filled with mercury whereas ground stress measurements are carried out with cells filled with oil. If stress measurements are to be monitored then ground load cells and concrete pressure cells should be grouped in pairs.

SEM Monitoring Cross Sections

Monitoring devices are grouped into monitoring cross sections (MCS). These MCS are depicted with their respective instrument layout indicating location and number of instruments within that MCS. Typical MCSs are shown on the design drawings for each individual tunnel cross-section geometry and excavation sequence.

Interpretation of Monitoring Results

All readings must be thoroughly and systematically collected and recorded. An experienced SEM tunnel engineer, often the SEM tunnel designer, must evaluate the data, occasionally complemented by visual observations of the initial shotcrete lining for any distress, for example as indicated by cracking. To establish a direct relationship between the behaviour of the tunnel and the ground as these react to tunnel excavation it is recommended to portray the development of monitoring values as a function of the tunnel progress. This involves a combined graph showing the monitoring value (i.e. deformation, stress or other) vs. time and the tunnel progress vs. time. In this example a prototypical deformation of a surface settlement point located above the tunnel centreline has been graphed on the ordinate (left vertical axis) vs. time on the horizontal axis.

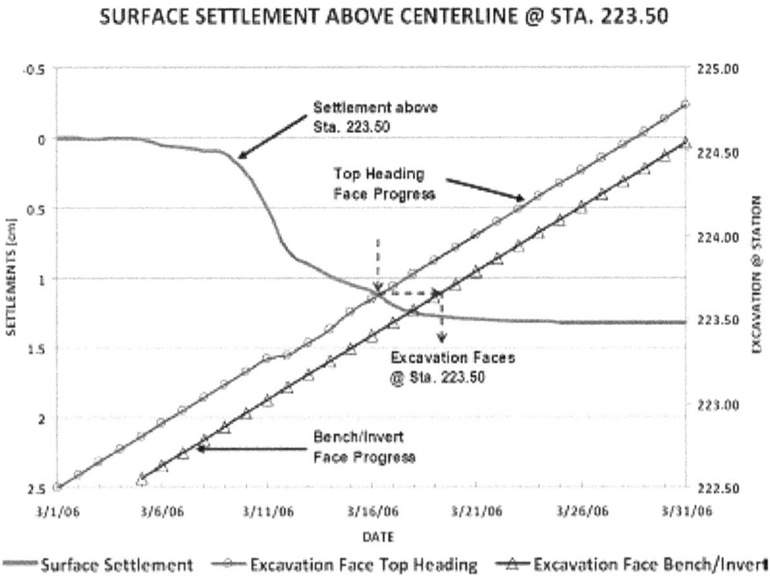

Figure: Prototypical Monitoring of a Surface Settlement Point Located Above the Tunnel Centreline in a Deformation vs. Time and Tunnel Advance vs. Time Combined Graph

The same time horizontal axis is used to portray the tunnel excavation progress by station location on the right vertical axis. As can be seen from this graph the surface settlement increases as the top heading and later bench/invert faces move towards and then directly under that point and gradually decrease as both faces again move away from the station location of the surface settlement point.

The settlement curve shows an asymptotic behaviour and becomes near horizontal as the faces are sufficiently far away from the monitoring point indicating that no further deformations associated with tunnel excavation and support occur in the ground indicating equilibrium and therefore ground stability.

The evaluation of monitoring results along with the knowledge of local ground conditions portrayed on systematic face mapping sheets forms the basis for the verification of the selected excavation and support class (ESC) or the need to make any adjustments to it.

Contractual Aspects

SEM construction requires solid past experience and personnel skill. This skill relates to the use of construction equipment and handling of materials for installation of the initial support including shotcrete, lattice girders, pre-support measures, and rock reinforcing elements and even more importantly observation and evaluation of the ground as it responds to tunnelling. It is therefore important to invoke a bidding process that addresses this need formally by addressing contractor qualifications and skills and payment on a unit price basis.

Contractor Pre-Qualification

It is recommended that the bidding contractors be pre-qualified to assure a skilled SEM tunnel execution. This pre-qualification can occur very early on during the design development but at a minimum should be performed as a separate step prior to soliciting tunnel bids. On critical SEM projects such as the NATM tunnelling at Russia Wharf in Boston in the late 90's the project owner solicited qualifications from contractors at the preliminary design stage.

This pre-qualification resulted in a set of pre-qualified contractors that were invited to comment on the design at the preliminary and intermediate design stages. This early process ensured that contractors were aware of the upcoming work and could plan ahead in assembling a qualified work force.

Pre-qualification documents shall identify the scope of work and call for a similar experience gained on past projects by the tunnelling company and key tunnelling staff including project manager, tunnel engineers, and tunnel superintendents. As a minimum the documents lay out description of ground conditions, tunnel size and length, excavation and support cycles, and any special methods intended for ground improvement.

Unit Prices

It is recommended that SEM tunnelling be procured within a unit price based contract. Unit prices suit the observational character of SEM tunnelling and the need to install initial support in accordance with a classification system and amount of any additional initial or local support as required by field conditions actually encountered. The following shall be bid on a unit price basis:

- Excavation and Support on a linear foot basis for all excavation and installation of initial support per Excavation and Support Class (ESC). This shall include any auxiliary measures needed for dewatering and ground water control at the face.
- Local support measures including:
  - o Shotcrete per cubic yard installed.
  - o Pre-support measures such as spiling, canopy pipes and any other support means such as rock bolts and dowels, lattice girders, and face dowels shall be paid per each (EA) installed.
  - o Instrumentation and monitoring shall be paid for either typical instrument section (including all instruments) or per each instrument installed. Payment will be inclusive of submitted monitoring results and their interpretation.
  - o Ground improvement measures per unit implemented, for example amount of grout injected including all labour and equipment utilised.

Waterproofing and final lining installed to complete the typical SEM dual lining structure may be procured on either lump sum basis or on a per tunnel foot basis.

The quantity of local support (additional initial support) measures shall be part of the contract to establish a basis for bid.

Experienced Personnel in Design, Construction, and Construction Management

Because the SEM relies on tunnelling experience it is imperative that experienced personnel be assigned from the start of the project, i.e., in its planning and design phase. The SEM design must be executed by an experienced designer. At this level of project development it is incumbent upon the owner to select a team that includes a tunnel designer with previous, proven, and relevant SEM tunnelling design and construction experience.

The SEM tunnel contract documents have to identify minimum contractor qualifications. In this case it is secondary whether the project is executed in a design-bid-build, design-build or any other contractual framework. For example, if the project uses the design-build framework then it is imperative that the builder take on an experienced SEM tunnel designer.

The construction contract documents must spell out minimum qualifications for the contractor's personnel that will initially prepare and then execute the SEM tunnel work. This is the case for field engineering, field supervisory roles and the labour force who must be skilled. SEM contract documents call for a minimum experience of key tunnelling staff by number of years spent in the field on SEM projects of similar type. Experienced personnel will include Senior SEM Tunnel Engineers, Tunnel Superintendents and Tunnel Foremen. All of such personnel should have a minimum of ten (10) years SEM tunnelling experience. These personnel are charged with guiding excavation and support installation meeting the key requirements of SEM tunnelling:

- Observation of the ground
- Evaluation of ground behaviour as it responds to the excavation process
- Implementation of the "right" initial support.

Knowledgeable face mapping, execution of the instrumentation and monitoring program, and interpretation of the monitoring results aid in the correct application of excavation sequencing and support installation. A typical face mapping form sheet that is used to document geologic conditions encountered in the field. While this form sheet portrays mapping for rock tunnelling, mapping of soft ground conditions is similar and lays out the characteristics of anticipated soil conditions. Face mapping should occur for every excavation round and be formally documented and signed off by both the contractor and the owner's representative.

The Senior SEM Tunnel Engineer is generally the contractor's highest SEM authority and supervises the excavation and installation of the initial support, installation of any local or additional initial support measures and pre-support measures in line with the contract requirements and as adjusted to the ground conditions encountered in the field. As a result the ground encountered is categorized in accordance with the contract documents into ground response classes (GRCs) and the appropriate excavation and support classes (ESC) per contract baseline. Any need for additional initial support and/or pre-support measures is assessed and implemented. This task is carried out on a daily basis directly at the active tunnel face and is discussed with the Owner's representative for each round. The outcome of this process is subsequently documented on form sheets that are then signed by the Contractor's and Owner's representatives for concurrence.

This frequent assessment of ground conditions provides for a continuous awareness of tunnelling conditions, for an early evaluation of adequacy of support measures and as needed for implementation of contingency measures that may involve more than additional initial support means. Such contingency measures may include heavy pre-support and face stabilisation measures or even systematic ground improvement measures.

To be able to support this on-going evaluation process on the owner's behalf the construction management (CM) and inspection team must also include SEM experience. These CM supervisory personnel are independent of the executing party and it is recommended that it include a designer's representative. However, it is often the case that the CM role is filled by a construction management entity that has been assigned an overall role for a project of which the tunnelling may only be a subset of the work. If this is the case it is important that the CM be thoroughly familiar with the SEM tunnel design and its design basis. For this purpose it is recommended that the CM participate in the design review process during design development from an early stage through the bidding of the tunnel work. If it is not possible to integrate the tunnel designer within the CM staff then the CM should be augmented by third party SEM experienced personnel who then oversee the tunnel execution in the field. The key of safe and successful SEM tunnelling is a solid knowledge of SEM principles and thorough experience with its execution.

Chapter 7

Tunnel Lining

The structural design, detailing and construction of tunnel linings for highway tunnels focusing on mined or bored tunnels. Tunnel linings are structural systems installed after excavation to provide ground support, to maintain the tunnel opening, to limit the inflow of ground water, to support appurtenances and to provide a base for the final finished exposed surface of the tunnel. Tunnel linings can be used for initial stabilisation of the excavation, permanent ground support or a combination of both. The materials for tunnel linings covered in this chapter are cast-in-place concrete lining, precast segmental concrete lining, steel plate linings, and shotcrete lining.

Figure: Cumberland Gap Tunnel

Cast-in-place concrete linings are generally installed some time after the initial ground support. Cast-in-place concrete linings are used in both soft ground and hard rock tunnels and can be constructed of either reinforced or plain concrete. Cast-in-place concrete linings can take on any geometric shape, with the shape being determined by the use, mining method and ground conditions.

Precast concrete linings are used as both initial and final ground support. Segments in the shape of circular arcs are precast and assembled inside the shield of a tunnel boring machine to form a ring. If necessary they can be used in a two pass system as only the initial ground support. Initial Support Segments for a two pass system are often lightly reinforced and rough cast. The second pass or final lining typically is cast-in-place concrete. Precast concrete linings can also be used in a one pass system where the segments provide both the initial and final ground support. One pass precast segmental concrete linings are cast to strict tolerances and are provided with gaskets and bolted together to reduce the inflow of water into the tunnel.

Figure: Precast Segmental Lining

Steel plate linings (liner plates) are a type of segmental construction where steel plates are fabricated into arcs that typically are assembled inside the shield of a tunnel boring machine to form a ring. The steel plate lining may form the initial and final ground support. The segments are provided with gaskets to limit the inflow of ground water into the tunnel. Steel plates are also used in lieu of lagging where steel ribs are used as the initial ground support. With the advent of precast concrete segments, liner plates are not used as much as previously.

Shotcrete is a pneumatically applied concrete that is used frequently as an initial support but now with the advances in shotcrete technology permanent shotcrete lining is designed and constructed in conjunction with sequential excavation method (SEM) tunnelling. One of the first applications of final shotcrete lining in the United States is at Lehigh Tunnel No. 2 of Pennsylvania Turnpike. Shotcrete can take on a variety of compositions. It can be applied over the exposed ground, reinforcing steel, welded wire fabric or lattice girders. It can be used in conjunction with rock bolts and dowels, it can contain steel or plastic fibers and it can be composed of a variety of mixes. It is applied in layers to achieve the desired thickness.

Figure: Baltimore Metro Steel Plate Lining

Cross passages and refuge areas are usually mined by hand after the main tunnel is excavated. These areas, due to their unique shape and small areas are typically lined with cast-in-place concrete. There is insufficient quantity involved in the lining of these features to make prefabricated linings economic.

Load and Resistance Factor Design (LRFD)

The design of tunnel linings, with the exception of steel tunnel lining plates, is not addressed in standard design codes. This chapter is intended to establish procedures for the design of tunnel linings utilising the American Association of State Highways and Transportation Officials (AASHTO) LRFD Bridge Design Specifications, current edition. LRFD is a design philosophy that takes into account the variability in the prediction of loads and the variability in the behaviour of structural elements. It is an extension of the load factor design methodology that has been in use for a number of years. This chapter is intended to assist the designer in the application of the LRFD specifications to tunnel lining design and to provide for a uniform interpretation of the AASHTO LRFD specification as it applies to tunnel linings.

Design Considerations

Lining Stiffness and Deformation

Tunnel linings are structural systems, but differ from other structural systems in that their interaction with the surrounding ground is an integral aspect of their behaviour, stability and overall load carrying capacity. The loss or lack of the support provided by the surrounding ground can lead to failure of the lining. The ability of the lining to deform under load is a function of the relative stiffnesses of the lining and the surrounding ground. Frequently, a tunnel lining is more flexible than the surrounding ground. This flexibility allows the lining to deform as the surrounding ground deforms during and after tunnel excavation. This deformation allows the surrounding ground to mobilise strength and stabilise. The tunnel lining deformation allows the moments in the tunnel lining to redistribute such that the main load inside the lining is thrust or axial load. The most efficient tunnel lining is one that has high flexibility and ductility.

A tunnel lining maintains its stability and load carrying capacity through contact with the surrounding ground. As load is applied to one portion of the lining, the lining begins to deform and in so doing, develops passive pressure along other portions of the lining. This passive pressure prevents the lining from buckling or collapsing. Ductility in the lining allows for the creation of "hinges" at points of high moment that relieve the moments so that the primary load action is axial force. This ductility is provided for in concrete by the formation of cracks in the concrete. Under reinforcing or no reinforcing help promote the initiation of the cracks. The joints in segmental concrete linings also provide ductility. In steel plate linings, the negligible bending stiffness of the steel plates and the inherent ductility of steel allow for the creation of similar hinges.

Constructability Issues

Each tunnel is unique. Ground conditions, tunnelling means and methods, loading conditions, tunnel dimensions and construction materials all vary from tunnel to tunnel. Each tunnel must be assessed on its own merits to identify issues that should be considered during design such that construction is feasible. Some common elements that should be considered are as follows:

Materials: Selection of tunnel lining materials should be made to facilitate transportation and handling of the materials in the limited space inside a tunnel. Pieces should be small and easily handled. Piece lengths should be checked to ensure that they can negotiate the horizontal and vertical geometry of the tunnel. Materials should be nontoxic and nonflammable.

Details: Detailing should be performed to facilitate ease of construction. For example, sloping construction joints in cast-in-place concrete linings can eliminate the difficulty associated with building a bulkhead against an irregular excavated surface.

Procedures: Construction procedures should be specified that are appropriate for conditions encountered in the tunnel; conditions that are often moist or wet, sometimes even with flowing water. Allow means and methods that do not block off portions of the tunnel for significant periods of time. The entire length of the tunnel should be available as much as practical.

Durability

Tunnels are expensive and are constructed for long term use. Many existing tunnels in the United States have been in use for well over one hundred years with no end in sight to their service lives. Having a tunnel out of service for an extended period of time can result in great economic loss. As such, details and materials should be selected that can withstand the conditions encountered in underground structures. All structures, including tunnels require inspection, periodic maintenance and repair tunnel inspection, maintenance, and rehabilitation. Nonetheless, detailing should be such that anticipated maintenance is simplified and long term durability is maximised.

Highway tunnels can also be exposed to extreme events such as fires resulting from incidents inside the tunnel. Tunnel lining design should consider the effects of a fire on the lining. The lining should be able to withstand the heat of the fire for some period of time without loss of structural integrity. The length of time required will be a function of the intensity of the anticipated fire and the response time for emergency personnel capable of fighting the fire. The tunnel lining should also sustain as little damage as possible so that the tunnel can go back into service as soon as possible. Protection from fire can be gained from concrete cover, tunnel finishes and the inclusion of plastic fibers in concrete mixes.

High Density Concrete

High density concrete is produced by using very finely ground cement and/or substituting various materials such as fly ash or blast furnace slag for cement. The cementitious content of high density concrete is very high. The high cement content makes handling difficult under ideal conditions. Complicated mixes with multiple admixtures and careful water monitoring are required to keep the concrete in a plastic state long enough to be placed in forms. High cement content will result in high heat of hydration. Proper curing of these materials is essential to produce a quality end product. Improper or incomplete curing can be the cause of severe cracking due to shrinkage. Shrinkage cracks can reduce the effectiveness of the product, affect its durability and potentially make it unusable.

High density concrete, however, can be beneficial in many tunnel applications. It can limit the inflow of water and provide significant protection against chemical attack. High density concrete has low heat conductivity which is beneficial in a fire. High density concrete should be used in conjunction with careful inspection and strict enforcement of specifications during construction.

Corrosion Protection

Corrosion is associated with steel products embedded in the concrete and otherwise used in tunnel applications. Ground water, ground chemicals, leaks, vehicular exhaust, dissimilar metals, deicing chemicals, wash water, detergents, iron eating bacteria and stray currents are all sources of corrosion in metals. Each of these and any other aspect that is unique to the tunnel under consideration must be evaluated during the design phase. Corrosion protection methods designed to combat the source of corrosion should be incorporated into the design.

Corrosion protection can take the form of coatings such as epoxies, powder coatings, paint or galvanizing. Insulation can be installed between dissimilar metals and sources of stray currents. High density concrete can provide protection for reinforcing steel. Coatings on concrete can minimise the infiltration of water, a component of almost all corrosion processes. Tunnel finishes can also protect the tunnel structural elements from attack by the various sources of corrosion.

Cathodic protection uses sacrificial material to protect the primary material from corrosion. In highly corrosive environments, an electrical current is induced in the materials to force corrosion to occur in the sacrificial material. These systems are highly effective when properly designed, installed and maintained. Sacrificial elements must be replaced and electrical supply equipment serviced regularly. Cathodic protection also requires a reliable long term source of electricity and adds to the maintenance and operation costs of the tunnel.

Increased concrete cover over reinforcing steel is an effective means of protecting reinforcing steel from corrosion. Increasing the concrete cover, however will also increase the thickness of the lining. The increased thickness will result in a larger excavation which will increase the overall cost of the tunnel. The use of increased concrete cover should be evaluated in terms of the overall cost of the tunnel compared to the benefit derived.

Lining Joints

Joints in linings are required to facilitate construction. Cast-in-place concrete requires construction joints. Construction joints can be sloped or formed. Segmental linings constructed from concrete or steel can have either bolted or unbolted joints. Unbolted joints are used in both gasketed and ungasketed concrete segments. Steel liner plates are bolted. More detailed information on the advantages and disadvantages of joints is provided in subsequent sections of this chapter.

Joints in linings also provide relief from stresses induced by movements due to temperature changes. Cast-in-place linings should have contraction joints every 30 feet and expansion joints every 120 feet. Expansion joints should also be used where cut and cover portions of the tunnel transition to the mined portion. Segmental concrete linings do not require contraction joints and require expansion joints only at the cut and cover interface.

Structural Design

Structural design will be governed by the latest AASHTO LRFD Bridge Design Specifications. The AASHTO specifications do not cover structural plain concrete which is frequently used in tunnel lining construction. This chapter will provide design procedures based on the AASHTO specifications for structural plain concrete.

Structural Analysis

Structural analysis of tunnel linings has been a subject of numerous papers and theories. Great disparity of opinion exists on the accuracy and usefulness of these analyses. However, some rational method must be adopted to determine a lining's ability to maintain the excavated opening of a tunnel. Some widely accepted methods are described in this section.

Beam Spring Models: A general purpose structural analysis program can be used to model the soil structure interaction. This method is known as the beam spring model. The computer model is constructed by placing a joint or node at points along the centroid of the lining. These nodes are joined by straight beam members that approximate the lining shape by a series of chords. When constructing this type of model, the chord lengths should be approximately the same as the lining thickness for the radii that can be expected in highway tunnels. Chord members that are too long can produce fictitious moments and chord members that are too short can result in computational difficulties because of the very small angles subtended by short members. A subtended angle dimension of approximately 60/R, where R is the radius of the tunnel in feet, will generally produce acceptable results. Properties such as cross sectional area and moment of inertia should be entered to accurately depict the real behaviour of the lining. Since the compressive forces are generally large enough to have compression over the entire thickness of the lining, the area and moment of inertia are calculated using the gross, uncracked dimensions of the lining. In rock tunnels, overbreak will result in a lining thickness larger than the design thickness. The design thickness is used in the analysis. This type of model is useful in analysing all geometric shapes.

The surrounding ground is modelled by placing a spring support at each joint. Springs can be placed in the radial and tangential directions. The tangential springs offer little value in the analysis and an unnecessary complication to the model. The numerical value of the spring constant at each support is calculated from the modulus of subgrade reaction of the surrounding ground multiplied by the tributary length of lining on each side of the spring. Many ground conditions can be encountered within the length of a single tunnel. Parametric studies that vary the ground conditions and the spring constants should be performed to determine the worst case scenario for the lining. Loads are applied to the model and the displacement at each joint is checked. For joints that move away from the centre of the tunnel into the ground, the spring is left active. When the joint displacement is toward the centre of the tunnel, the spring is removed or made inactive. This process in repeated until all displacements match the spring condition (active or inactive) at that joint. Once the model converges, the moments, thrusts and shears are used to design the lining.

If the model reveals that the lining is beyond its capacity, making the lining thicker or stiffer will not alleviate the problem. In fact, stiffening the lining will cause it to attract more moment and it will likely continue to fail. The lining must be made to be more flexible. This can be accomplished by making the lining thinner, which may not work. The primary load action on the lining is axial load or thrust. If the lining is close to its capacity under this load action, then thinning will not work. Modelling lining flexibility such that the moments are relieved may show the lining to be adequate. This is what happens in reality. One way to model this phenomenon is to install full or partial hinges in the lining at points of theoretical high moment. The hinge can be modelled to accept as much moment as the lining can support or it can be modelled as a full hinge with no moment capacity. In reality, the lining is performing somewhere in between these two extremes. Analysing both conditions will bracket the lining behaviour and provide a reasonable assurance that the lining can support the loads.

Three Dimensional Models: The model described above is usually a two dimensional model that represents a single foot along the length of the tunnel. More sophisticated models are required when large penetrations of the lining or intersecting tunnels are being analysed. To model these conditions, a three dimensional finite element model is used. The model is constructed in a similar manner to the two dimensional model, with finite elements used to connect the nodes and create the three dimensional model. The modelling parameters described above hold true for this type of model also. The model should extend a minimum of one tunnel diametre beyond the feature being investigated on each side of the feature.

It has been argued that this model does not account for the nonlinearity of the surrounding ground, particularly in soft ground, nor does it account for the variation of ground movement with time. Careful development of loading diagrams and spring constants for this model can bracket the actual behaviour of the surrounding ground. This will provide results that are comparable to more sophisticated analysis methods. It should be noted that this method of analysis typically over estimates the bending moment in the lining.

Empirical Method for Soft Ground: For circular tunnels in soft ground, the validity of the beam spring model has been highly criticized. The beam spring model described above assumes the soil to be a homogenous elastic material when in fact it is often non-homogenous and the behaviour is plastic rather than elastic. Plastic deformations of the soil take place and the lining "goes along for the ride", that is, the stiffness of the lining is incapable of resisting the soil deformations. Since the lining is typically more flexible than the surrounding soil, it distorts as the soil displaces and the lining's flexibility allows it to shed moments to the point where it is acting almost entirely in compression. Since the lining is not completely flexible, some residual moment remains in the lining. This moment is accounted for by assigning an arbitrary change in radius and calculating the theoretical moment resulting from this change in radius. Using this method, the thrust in the tunnel lining is calculated by the formula:

$$T = wR$$

Where:

T = the thrust in the tunnel lining

w = the earth pressure at the spring line of the tunnel due to all load sources

R = the radius of the tunnel

The percentage of radius change to be used is a function of the type of soil.

Table 1: Percentage of Lining Radius Change in Soil

Soil Type	ΔR/R - Range
Stiff to Hard Clays	0.15 - 0.40%
Soft Clays or Silts	0.25 - 0.75%
Dense or Cohesive Soils, Most Residual Soils	0.05 - 0.25%
Loose Sands	0.10 - 0.35%

Notes:

1. Add 0.1 to 0.3 percent for tunnels in compressed air, depending on air pressure.
2. Add appropriate distortion for effects such as passing neighbour tunnel.
3. Values assume reasonable care in construction, and standard excavation and lining methods.

The resulting bending moment in the lining is calculated using the following formula:

$$M = 3EI/R \times \Delta R/R$$

Where:

M = the calculated bending moment

R = radius to the centroid of the lining

ΔR = tunnel radius change

E = modulus of elasticity of the lining material

I = effective moment of inertia of the lining section

The effective moment of inertia can be calculated for precast segmental linings using the following formula:

$$I_e = I_j + I(4/n)^2$$

Where:

I_e = The effective moment of inertia

I_j = The joint moment of inertia (conservative taken as zero)

I = The moment of inertia of the gross lining section

n = The number of joints in the lining ring

The moment of inertia for the uncracked section should be used for cast-in-place concrete linings.

Numerical Methods Commercial software is also available to model both the lining and the surrounding ground as a continuum utilising a three dimensional finite element or finite difference approach. FLAC3D is a finite difference based continuum analysis program, where the domain (ground) is assumed to be a homogeneous media. The structural elements (beam or shell elements) can be used to model the tunnel lining. Using interface elements between the lining elements and the surrounding ground, rock-lining interaction including slip can be simulated.

If the ground contains predominant weak planes and those are continuous and oriented unfavourably to the excavation, then the analysis should consider incorporating specific characteristics of these weak planes.

In this case, mechanical stiffness (force/displacement characteristics) of the discontinuities may be much different from those of intact rock. Then, a discrete element method (DEM) can be considered to solve this type of problem.

3DEC is a commercially available program for this type of analysis. Unlike continuum analysis, the DEM permits a large deformation and finite strain analysis of an ensemble of deformable (or rigid) bodies (intact rock blocks) which interact through deformable, frictional contacts (rock joints).

It is greatly task dependent whether a continuum (FLAC3D) or discrete analysis (3DEC) is adequate. If the ground is soil, the FLAC3D is adequate. If the ground is jointed rockmass and the joints are predominant in rock-lining interaction, 3DEC should be utilised. These programs can be used to calibrate and verify beam spring models, and vice versa.

Cast-in-Place Concrete

Cast-in-place concrete linings are used as final linings in two pass lining systems. Initial ground support is installed in the tunnel as the tunnel is excavated and can take any form from steel ribs and lagging to precast concrete segments. A water proofing system or drainage blanket is typically placed between the initial ground support and the cast-in-place concrete lining.

The typical section for the cast-in-place lining used for the Cumberland Gap Tunnel. The Cumberland Gap Tunnel is a highway tunnel excavated in rock by the drill and blast method. Initial ground support is untreated rock, shotcrete and rock bolts. The initial ground support varied along the length of the tunnel due to varying ground conditions.

Advantages of a cast-in-place concrete lining are as follows:

- Suitable for use with any excavation and initial ground support method.
- Corrects irregularities in the excavation.
- Can be constructed to any shape.
- Provides a regular sound foundation for tunnel finishes.
- Provides a durable low, maintenance structure.

Figure: Cast-In-Place Concrete Lining, Washington DC

Disadvantages of a cast-in-place concrete lining are as follows:

- Concrete placement, especially around reinforcement can be difficult. The nature of the construction of the lining restricts the ability to vibrate the concrete. This can result in incomplete consolidation of the concrete around the reinforcing steel.

- Reinforcement when used is subject to corrosion and resulting deterioration of the concrete. This is a problem common to all concrete structures, however underground structures can be also be subject to corrosive chemicals in the groundwater that could potentially accelerate the deterioration of reinforcing steel.
- Cracking that allows water infiltration can reduce the life of the lining.
- Chemical attack in certain soils can reduce lining life.
- Construction requires a second operation after excavation to complete the lining.

Design Considerations

In order to maximise flexibility and ductility, a cast-in-place concrete lining should be as thin as possible. There are, however, practical limits on how thin a section can be placed and still obtain proper consolidation and completely fill the forms. 10 inches (25 cm) is considered the practical minimum thickness for a cast-in-place concrete lining.

Reinforcing steel in a thin section can also be problematic. The reinforcement inhibits the flow of the concrete making it more difficult to consolidate. If two layers of reinforcement are used, then staggering the bars may be required to obtain the required concrete cover over the bars. This can make the forms congested and concrete placement more difficult. Self consolidating concrete has been in development in recent years and has been used in unreinforced concrete linings in Europe with some success. Self consolidating concrete may prove useful in reinforced concrete linings, however it recommended that an extensive testing program be made part of the construction requirements to ensure that proper results are, in fact, obtained.

Cast-in-place concrete is used as the final lining. In many cases a waterproofing system is placed over the initial ground support prior to placing the final concrete lining. Placing reinforcing steel over the waterproofing system increases the potential for damaging the waterproofing. In all cases that are practical to do so, cast-in-place concrete linings should be designed and constructed as plain concrete, that is with no reinforcing steel. The presence of the waterproofing systems precludes load sharing between the final lining and the initial ground support. A basic design assumption is that the final lining carries long term earth loads with no contribution from the initial ground support.

Ground water chemistry should be investigated to ensure that chemical attack of the concrete lining will not occur should the lining be exposed to ground water. If this is an issue on a project, mitigation measures should be put in place to mitigate the effects of chemical attack. The waterproofing membrane can provide some protection against this problem. Admixtures, sulfate resistant cement and high density concrete may all be potential solutions. This problem should be addressed on a case by case basis and the appropriate solution be implemented based on best industry practice.

Concrete behaviour in a fire event must also be considered. When heated to a high enough temperature, concrete will spall explosively. This produces a hazardous condition for motorists attempting to exit the tunnel and for emergency response personnel responding to the incident. This spalling is caused by the vapourization of water trapped in the concrete pores being unable to escape. Spalling is also caused by fracture of aggregate and loss of strength of the concrete matrix at the surface of the concrete after prolonged exposure to high temperatures. Reinforcing steel that is heated will loose strength. Spalling and loss of reinforcing strength can cause changes in the shape of the lining, redistribution of stresses in the lining and possibly structural failure.

The lining should be protected against fire. Both external and internal protection can be provided. External protection in the form of coatings or boarding is available commercially. These are speciality products that can provide a measure of protection against relatively low temperature fires. Manufacturers should be consulted to ascertain the exact level of protection that they can provide. Including polypropylene fibers in the concrete mix can reduce vapourization of entrapped water. The fibers melt during a fire and provide a pathway for water to escape.

Materials

Mixes for cast-in-place concrete should be specified to have a high enough slump to make placement practical. A slump of 5" (12.7 cm) is recommended. Air entrainment should be used. The moist environment in many tunnels combined with exposure to cold weather makes air entrainment important to durable concrete; 3 to 5 percent air entrainment is recommended.

Compressive strength should be kept to a minimum. High strength concretes require complex mixes with multiple admixtures and special placing and curing procedures. Since concrete lining acts primarily in compression, 28 day compressive strengths in the range of 3,500 to 4,500 psi (24 to 31 MPa) are generally adequate.

Reinforcing steel bars should conform to the requirements of ASTM A615 grade 60 and welded wire fabric when used should conform to ASTM A185.

Construction Considerations

Cast-in-place concrete must attain a minimum strength prior to stripping forms. The concrete must also be cured. Leaving the forms in place can accomplish both these goals, but can inhibit the rate of construction. The concrete should reach some minimum strength prior to stripping the forms. This should be computed by the designer assuming that the tunnel is supported by the initial support and thus the final lining at the time of stripping will be carrying only its own weight. The strength of the concrete in the forms can be verified by breaking field cured cylinders. This will allow the forms to be stripped as soon as possible. Curing can continue after stripping by keeping the concrete moist or by applying a curing compound. Curing compounds should only be used if the concrete is the finished exposed surface. The curing compound will act as a bond breaker if finishes such as ceramic tile are applied to the concrete. Sealants and coating will not adhere to concrete surfaces that have had curing compound applied unless the curing compound is removed via sand blasting or other technique.

The length of pour along the centreline of tunnel should be limited to minimise shrinkage in the concrete. Lining forms are usually designed to be re-used so limiting the length of pour does not impose a hardship on the contractor. Construction joints can be bulkheaded or sloping. Bulkheaded joints provide a uniform appearance, however, depending on how uneven the face of excavation is, construction of the bulkhead may be difficult. Sloped construction joints do not affect the performance of the lining, but can be unattractive and should be rubbed out after the forms are stripped.

Placing concrete in a curved shape overhead will leave a void at the crown. This void is filled after the concrete is cured by pumping grout into the void. Grout pipes are installed in the forms prior to placing the concrete to facilitate this operation. Spacing of the grout pipes along the tunnel should be limited to 10 feet and the pipes should be offset from the crown by 15 degrees on both sides.

When appurtenances are attached to the finished concrete lining, epoxy type anchor bolts should never be used. It is recommended to use undercut mechanical anchors to attach appurtenances to tunnel linings.

Precast Segmental Lining

Precast segmental linings are used in circular tunnels that are mined using a tunnel boring machine. They can be used in both soft and hard ground. Several curved precast elements or segments are assembled inside the tail of the tunnel boring machine to form a complete circle. The number of segments used to form the ring is a function of the ring diametre and to a certain respect, contractor's preferences. The segments are relatively thin, 8 to 12 inches (20 to 30 cm) and typically 40 to 60 (1 to 1.5 m)inches (cm) wide measured along the length of the tunnel.

Precast segmental linings can be used as initial ground support followed by a cast-in-place concrete lining (the "two-pass" system) or can serve as both the initial ground support and final lining (the "one-pass" system) straight out of the tail of the TBM. Segments used as initial linings are generally lightly reinforced, erected without bolting them together and have no waterproofing. The segments are erected inside the tail of the TBM. The TBM pushes against the segments to advance the tunnel excavation. Once the shield of the TBM is passed the completed ring, the ring is jacked apart (expanded) at the crown or near the springlines. Jacking the segments helps fill the annular space that was occupied by the shield of the TBM. After jacking, contact grouting may be used to finish filling the annular space and to ensure complete contact between the segments and the surrounding ground. A waterproofing membrane is installed over the initial lining and the final concrete lining is cast in place against the waterproofing membrane. Horizontal and vertical curvature in the tunnel alignment is created by using tapered rings. The curvature is approximated by a series of short chords.

Precast segmental linings used as both initial support and final lining are built to high tolerances and quality. They are typically heavily reinforced, fitted with gaskets on all faces for waterproofing and bolted together to compress the gaskets after the ring is completed but prior to advancing the TBM. As the completed ring leaves the tail of the shield of the TBM, contact grouting is performed to fill the annular space that was occupied by the shield. This provides continuous contact between the ring and the surrounding ground and prevents the ring from dropping into the annular space. Bolting is often performed only in the circumferential direction. The shove of the TBM is usually sufficient to compress the gaskets in the longitudinal direction. Friction between the ground and the segments hold the segment in place, maintaining compression on the gasket. When first introduced into the United States in the mid-1970's, segmental linings were fabricated in a honeycomb shape that allowed for bolting in both the longitudinal and circumferential directions.

The lining Baltimore Metro. After 30 years of service, this lining is still providing a stable dry opening for over a hundred trains per day. Recent lining designs have eliminated the longitudinal bolting and the complex forming and reinforcing patterns that were required to accommodate the longitudinal bolts. Segments now have a flat inside surface as shown. The segments in the casting bay after being stripped of the forms. Once adequate strength is achieved, the segments are inverted to the position they must be in for erection inside the side the tunnel. Segments are generally stored in a stacked arrangement, with one stack containing the segments required to construct a single ring inside the tunnel. As with segments used for initial lining, horizontal and vertical tunnel alignment is achieved through the use of tapered segments.

Advantages of a precast segmental lining are as follows:

- Provides complete stable ground support that is ready for follow-on work.
- Materials are easily transported and handled inside the tunnel.
- No additional work such as forming and curing is required prior to use.
- Provides a regular sound foundation for tunnel finishes.
- Provides a durable low maintenance structure.
- Disadvantages of a precast segmental lining are as follows:
- Segments must be fabricated to very tight tolerances
- Reinforcing steel must be fabricated and placed to very tight tolerances.
- Storage space for segments is required at the job site.
- Segments can be damaged if mishandled.
- Spalls, cracked and damaged edges can result from mishandling and over jacking.
- Gasketed segments must be installed to high tolerances to assure that gaskets perform as designed.

- Reinforcement when used is subject to corrosion and resulting deterioration of the concrete.
- Cracking that allows water infiltration can reduce the life of the lining.
- Chemical attack in certain soils can reduce lining life.

Figure: Precast Segments for One-Pass Lining, Forms Stripped

Figure: Stacked Precast Segments for One-Pass Lining

Design Considerations

Initial Lining Segments used as an initial support lining are frequently designed as structural plain concrete. Reinforcing steel is placed in the segments to assist in resisting the handling and storage loads imposed on the segments. Reinforcement is often welded wire fabric or small reinforcing steel bars. The segments are usually cast by a precaster or in a yard set up specifically for manufacturing the segments.

Figure: Stacked Precast Segments for Two-Pass Lining

Figure: Steel Cage for Precast Segments for Two-Pass Lining

These segments are used as the initial lining and are not required to be waterproof. Therefore no gaskets are used. No keyway for a gasket in cast into the segment. Note however, the keyway cast into the sides of the segments used to help with placement of the segment and maintaining alignment of the segments in the radial direction. Figure shows the reinforcing steel cages for the segments.

When using a structural analysis program for analysis, the structural model should include hinges (points where no bending moment can develop) at the locations of the joints in the ring. Using hinges at the joint locations provides the ring with the flexibility required to adjust to the loads, resulting in the predominant loading being axial load or thrust. This is an approximation of the behaviour of the lining since joints will transfer some moment. The actual behaviour of a segmental lining can be bounded by models that have zero fixity at the joints and full fixity at the joints.

Radial joints in between segments can be flat or concave/convex as shown, Convex/concave joints facilitate rotation at the joint, allowing the segment to deform and dissipate moments. Flat joints are more efficient at transferring axial load between segments and may result in less end reinforcement. In either case, the ends of the segments that form the joints should be reinforced to facilitate the transfer of load from one segment to another without cracking and spalling. The amount of reinforcement used should consider the type of joint and the resulting load transfer mechanism. Handling and erecting the segments are also sources of damage at the joints. Reinforcing can mitigate this damage.

The primary load carried by the precast segments is axial load induced by ground forces acting on the circumference of the ring. However, loads imposed during construction must also be accounted for in the design. Loads from the jacking forces of the TBM are significant and can cause segments to be damaged and require replacement. These forces are unique to each tunnel and are a function of the ground type and the operational characteristics of the TBM. Reinforcement along the jacking edges of the segments is usually required to resist this force. The segments should be checked for bearing, compression and buckling from TBM thrust loads.

Handling, storage, lifting and erecting the segments also impose loads. The segments should be designed and reinforced to resist these loads. The dead weight of the segment with a dynamic factor of 2.0 applied to that dead weight is recommended for design to resist these loads. When designing reinforcement for these loads, the provisions of the LRFD specifications should be used. Grouting pressure can also impose loads on the lining. Grouting pressures should be limited to reduce the possibility of damage to the ring by these loads. A value of 10 psi (69 kPa) is recommended as the maximum permissible grouting pressure. The anticipated grouting pressure should be added to the load effects of the ground loads applied to the lining.

Initial lining segments are considered to be temporary support, therefore long term durability is not considered in the design of the linings or materials used.

Final Lining Segments used as a final lining are designed as reinforced concrete. The reinforcement assists in resisting the loads and limits cracking in the segment. Limiting cracking helps make the segments waterproof. The provisions of the AASHTO LRFD specifications should be used to design the segments. The segments are manufactured by a precaster or in a yard set up specifically for manufacturing the segments. Since the segments are cast and cured in a controlled environment, higher tolerances can be attained than in cast-in-place concrete construction.

When using a structural analysis program for analysis, an effective moment of inertia should be used to account for the flexibility induced in the ring at the bolted joints. The effective moment of inertia can be calculated using formula. When using this effective moment of inertia, no hinges are installed in the beam spring model.

Final lining segments can be fabricated with straight or skewed joints. A schematic of a lining system with straight joints. The orientation of the joint should be considered in the design of the lining to account for the mechanism of load transfer across the joint between segments. Skewed joints will induce strong axis bending in the ring and this should be accounted for in the design of the ring. Whether using straight or skewed joints, segments are rotated from ring to ring so that the joints do not line up along the longitudinal axis of the tunnel.

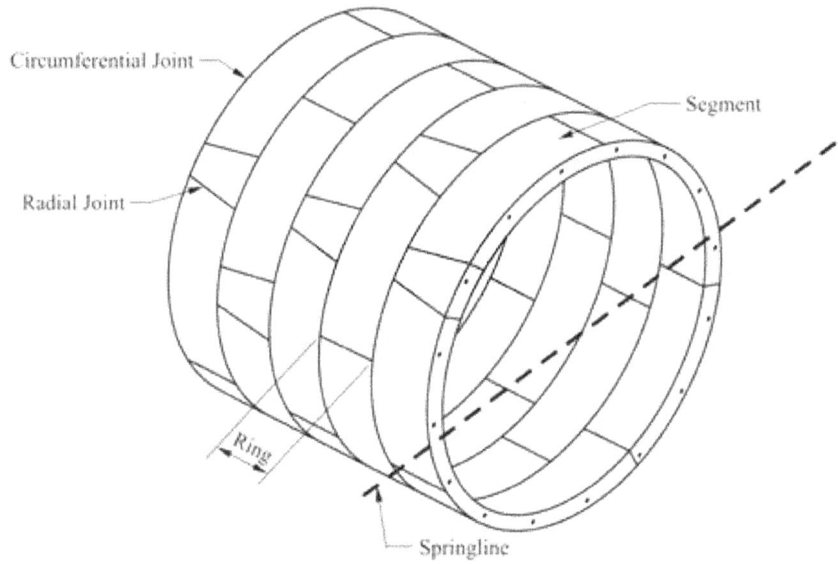

Figure: Schematic Precast Segment Rings

Joint design should consider the configuration of the gaskets. The gasket can eliminate much of the bearing area for load transfer between joints. Joints should be adequately reinforced to transfer load across the joints without damage.

The primary load carried by the precast segments is axial load induced by ground, hydrostatic and other forces acting on the circumference of the ring. The presence of the waterproofing systems precludes load sharing between the final lining and the initial ground support. A basic design assumption is that the final lining carries long term earth loads with no contribution from the initial ground support. Loads imposed during construction must also be accounted for in the design. Loads from the jacking forces of the TBM are significant and can cause segments to be damaged and require replacement. These forces are unique to each tunnel and are a function of the ground type and the operational characteristics of the TBM. Reinforcement along the jacking edges of the segments may be required to resist this force. The segments should be checked for bearing, compression and buckling from TBM thrust loads.

Lifting and erecting the segments also impose loads. The segments should be designed and reinforced to resist these loads. The dead weight of the segment with a dynamic factor of 2.0 applied to the dead weight is recommended for design to resist these loads. When designing reinforcement for these loads.

Grouting pressure can also impose loads on the lining. Grouting pressures should be limited to reduce the possibility of damage to the ring by these loads. A value of 10 psi (69 kPa) is recommended as the maximum permissible grouting pressure. The anticipated grouting pressure should be added to the load effects of the earliest ground loads applied to the lining.

Ground water chemistry should be investigated to ensure that chemical attack of the concrete lining will not occur should it be exposed to ground water. If this is an issue on a project, mitigation measures should be put in place to reduce the effects of chemical attack. The waterproofing membrane can provide some protection against this problem. Admixtures, sulfate resistant cement and high density concrete may all be potential solutions. This problem should be addressed on a case by case basis and the appropriate solution implemented based on best industry practice.

Concrete behaviour in a fire event must also be considered. When heated to a high enough temperature, concrete will spall explosively. This produces a hazardous condition for motorists attempting to exit the tunnel and to emergency response personnel responding to the incident. This spalling is caused by the vapourization of water trapped in the concrete pores being unable to escape. Spalling is also caused by fracture of aggregate and loss of strength of the concrete matrix at the surface of the concrete after prolonged exposure to high temperatures. Reinforcing steel that is heated will loose strength. Spalling and loss of reinforcing strength can cause changes in the shape of the lining, redistribution of stresses in the lining and possibly structural failure. The lining should be protected against fire. Both external and internal protection can be provided. External protection in the form of coatings or boarding is available commercially. These items can provide a measure of protection against relatively low temperature fires. These are speciality products and manufacturers should be consulted to ascertain the exact level of protection that they can provide. Including polypropylene fibers in the concrete mix can reduce vapourization of entrapped water. The fibers melt during a fire and provide a pathway for water to escape.

Appendix G presents a calculation example to illustrate the design process for precast segmental lining.

Materials

Concrete mixes for precast segments for initial linings do not require special designs and can generally conform to the structural concrete mixes provided in most state standard construction specifications. Strengths in the range of 4,000 to 5,000 psi (27 to 35 MPa) are generally adequate. These strengths are easily attainable in precast shops and casting yards. Curing is performed in enclosures and is well controlled. Air entrainment is desirable since segments may be stored outdoors for extended periods of time and final lining segments may be exposed to freezing temperatures inside the tunnel.

Steel fibre reinforced concrete has become a topic of discussion and research for precast tunnel linings. Theoretically, steel fibers can be used in lieu of steel reinforcing bars. The fibers can potentially eliminate the need for fabricating the steel bars to very tight tolerances, provide ductility for the concrete and make the segments tougher and less damage prone during construction. Unfortunately, there is no US design code for the design of steel fibre reinforced concrete. Papers have been written that propose design methods and several European countries have developed design methods. The recommended practice until further research is conducted and design codes are developed is to use steel fibers in segments where the design is conducted as detailed in this manual and the lining is found to be adequate without reinforcing. The steel fibers then can be included in the concrete to improve handling characteristics during construction. A testing program is required by the specifications to have the contractor prove via field testing that the fibre reinforced segments can withstand the handling loads imposed during construction. The fibers then can be used lieu of reinforcement that would be installed to resist the handling loads.

Reinforcing steel bars should conform to the requirements of ASTM A615 grade 60 and welded wire fabric when used should conform to ASTM A185. Concrete mixes for one pass lining segments have strengths ranging from 5,000 psi to 7,000 psi (34 to 48 MPa). Higher strengths are easily obtainable in precast shops and assist in resisting handling and erection loads.

Construction Considerations

Initial Lining Segments Grout holes are required for contact grouting. Grout holes can also serve as lifting points for the segments. Locate the grout holes symmetrically so that the load to the lifting devices is evenly applied. Grout holes and lifting devices are usually designed by the contractor to loads and criteria specified by the designer. The construction industry is moving toward vacuum erection and handling equipment. This device does not rely on the grout holes to handle the segments. A device of this type can be seen. This devices relies on vacuum created between the segment face and the device to produce the reaction required to lift and erect the segments.

Segments should be cast and cured in accordance with the requirements of the standard specifications of the owner. In the absence of standard specifications, the requirements of the Precast Concrete Institute should be used to develop construction specifications for the precast segments. Segments should be stored in a manner that will not damage the segments. Support locations should be shown on the drawings and maximum stacking heights should be specified.

Segments should be detailed to facilitate jacking the rings at the crown or near springline after erection. Space for material to temporarily close off the gap to stop earth from coming into the tunnel is required. A means to jack the segments should be devised and the space remaining from the jacking should be backfilled with concrete and/or contact grouting to complete the ring. The ends of the segments that are used for jacking may require additional reinforcing or steel plates to protect them from the forces associated with jacking.

The number of segments for two pass systems is usually kept at a minimum, with the segments being slightly larger than for a one pass system and the joints in the rings will line up with joints in adjacent rings.

Final Lining Segments (One-Pass System) The same considerations as for initial lining segments apply to final lining segments. Final lining segments, however are not jacked at the crown after erection. Final lining segments must also be detailed to accommodate the gaskets required for waterproofing. Often, final lining segments also receive a waterproofing coating applied to the outside of the segment. This waterproofing coating should be a robust material such as coal tar epoxy since the segments slide along the shield as it advances and damage to the coating will occur.

Steel Plate Lining

Steel plate lining is a segmental lining system. It is sometimes used for circular tunnels in soft ground mined by TBM or other methods. Several curved steel elements or segments are assembled inside the tunnel or the TBM to form a complete circle. The segments are constructed from steel plates that are pressed into the required shape. The plates have flanges along all four edges. The flanges are used to bolt the segments together in the longitudinal and circumferential directions. Adjacent rings are rotated so that joints do not line up from ring to ring. The segments are fitted with gaskets along all the flanges that are compressed when the bolts are tightened. These gaskets are intended to provide waterproofing for the tunnel. Lining plate is manufactured in standard sizes and in widths of either 12" (25.4 cm) or 24" (50.8 cm). Only the radius changes to meet the requirements of the project.

Advantages of a steel plate lining are as follows:

- Provides complete stable ground support that is ready for follow-on work.
- Materials are easily transported and handled inside the tunnel.
- No additional work such as forming and curing is required prior to being ready to use.

Disadvantages of a steel lining plate are as follows:

- Thrust applied from TBM must be limited to the capacity of the plate.
- Steel is subject to corrosion in the damp environment usually encountered in a tunnel.
- Fire can cause the lining plate to buckle and/or fail.
- Cast-in-place concrete will be needed for fire protection.

Design Considerations

Design of steel plate linings should be in accordance with the AASHTO LRFD specifications. The required checks for the service condition are included in that chapter. Typically, the design parameters and minimum dimensional requirements are specified on the drawings. The information on the drawings is developed from the AASHTO LRFD requirements. Lining plate manufacturers have standard products that can be selected for use on a project. The contractor will provide a specific product intended for use on the project and supply computations illustrating that the product meets the minimum requirements shown on the drawings.

The steel plate lining must be designed to resist jacking loads imposed by a TBM. A jacking ring or some other method of distributing these loads to the plates must be utilised to avoid damaging the plates during tunnelling. Often, stiffeners are required at the centre of the plates to resist the jacking loads. These stiffeners along with the flanges at the edges of the plates resist the bulk of this jacking force. The stiffeners and flanges are designed as columns to resist the anticipated jacking loads. Design of steel lining plate linings must include other loads induced by construction activities. Lifting and erection stresses should be checked by the contractor. Curvature in horizontal and vertical alignments is accommodated with tapered segments just as with concrete segments. Steel plate linings should be protected against corrosion. The exterior surface can be protected by a coating such as coal tar epoxy. The interior can be protected with coatings such as paint or galvanizing, but the most effective protection is a layer of unreinforced concrete. This concrete layer provides protection against corrosion and against heat damage due to fires. The protective concrete layer is placed after completion of the mining operation to avoid damage that can be caused by the jacking or the shield. Gasket requirements for steel plate linings are similar to those for concrete segments. However, steel plate linings have far less surface area for gasket installation than do concrete segments.

Shotcrete Lining

Shotcrete represents a structurally and qualitatively equal alternative to cast-in-place concrete linings. Its surface appearance can be tailored to the desired project goals. It may remain a rough, sprayer type shotcrete finish, or may have a quality comparable to cast concrete when trowel finish is specified. Shotcrete as a final lining is typically utilised in combination with the initial shotcrete supports in SEM applications when the following conditions are encountered:

- The tunnels are relatively short in length and the cross section is relatively large and therefore investment in formwork is not warranted, i.e. tunnels of less than 400-600 feet (150-250 m) in length and larger than about 25-35 feet (8-11 m) in springline diametre.

- The access is difficult and staging of formwork installation and concrete delivery is problematic.
- The tunnel geometry is complex and customized formwork would be required. Tunnel intersections, as well as bifurcations qualify in this area. Bifurcations are associated with tunnel widenings and would otherwise be constructed in the form of a stepped lining configuration and increase cost of excavated material.

When shotcrete is utilised as a final lining in dual shotcrete lining applications it will be applied against a waterproofing membrane as presented. The lining thickness will be generally 10 to 12 inches (200 to 300 mm) or more and its application must be carried out in layers with a time lag between layer applications to allow for shotcrete setting and hardening. To ensure a final lining that behaves close to monolithically from a structural point of view it is important to limit the time lag between layer applications and assure that the shotcrete surface to which the next layer is applied is clean and free of any dust or dirt films that could create a de-bonding feature between the individual layers. It is typical to limit the application between the layers to 24 hours. Shotcrete final linings are applied onto a carrier system that is composed of lattice girders and welded wire fabric mounted to lattice girders toward the waterproofing membrane side. This carrier system also acts fully or partially as structural reinforcement of the finished lining. The remainder of the required structural reinforcing may be accomplished by rebars or mats or by steel or plastic fibers. The final shotcrete layer allows for the addition of micro poly propylene (PP) fibers that enhance fire resistance of the final lining.

Unlike the hydrostatic pressure of cast-in-place concrete during installation the shotcrete application does not develop pressures against the waterproofing membrane and the initial lining and therefore one must ensure that any gaps between waterproofing system and initial shotcrete lining and final shotcrete lining be filled with contact grout. As in final lining applications contact grout is accomplished with cementitious grouts but the grout takes are much higher. To assure a proper grouting around the entire lining circumference it is customary to use longitudinal grout hoses arranged radially around the perimetre.

A typical shotcrete final lining section with waterproofing system, welded wire fabric (WWF), lattice girder, grouting hoses for contact grouting and a final shotcrete layer with PP fibre addition.

Probably the most important factor that will influence the quality of the shotcrete final lining application is workmanship. While the skill of the shotcrete applying nozzlemen (by hand or robot) is at the core of this workmanship, it is important to address all aspects of the shotcreting process in a method statement. This method statement becomes the basis for the application procedures, and the applicator's and the supervision's Quality Assurance / Quality Control (QA/QC) program. Minimum requirements to be addressed in the method statement are as follows:

- Execution of Work (Installation of Reinforcement, Sequence of Operations, Spray Sections, Time Lag)
- Survey Control and Survey Method
- Mix Design and Specifications
- QA/QC Procedures and Forms ("Pour Cards")
- Testing (Type and Frequency)
- Qualifications of Personnel
- Grouting Procedures

General trends in tunnelling indicate that the application of shotcrete for final linings presents a viable alternative to traditional cast-in-place concrete construction. The product shotcrete fulfills cast-in-place concrete structural requirements. Design and engineering, as well as application procedures, can be planned such as to provide a high quality product. Excellence is needed in the application itself and must go hand-in-hand with quality assurance during application.

Ship Anchors

The effect of an anchor impacting the underwater tunnel structure directly or being dragged across the line of the tunnel structure should be considered. Either the tunnel structure should be designed to resist the full loading imposed by the design anchor system, or the backfill / armour system should be designed to mitigate the loading, in which case the tunnel structure should be designed for the demonstrable reduced load. Rupture of the waterproofing membrane should not occur. The design anchor should be selected as appropriate to shipping using or expected to use the waterway, based on the relevant section of Lloyd's Rules.

The penetration depth of a falling anchor through tunnel roof protection material should be estimated. The formulae given in CEB Bulletin d'Information No 187, August 1988, reproduced for reference below provide a good design method to calculate the anchor penetration depth in granular material:

Penetration Depth of a Falling Anchor through Granular Material:

$$x = 10N_{pdn}d_e$$

$$N_{pdn} = \sqrt{\frac{m_w}{E_r d_e^3}}.vi$$

$$d_e - \sqrt{\frac{4A}{\pi}}$$

$$A = 0.6 + 0.2\frac{m_e}{1000}$$

where:

x penetration depth (m)

N_{pen} penetration parameter

d_e equivalent diametre of striking area of anchor (m)

m_w mass of anchor reduced by the mass of the displaced water (kg)

m_a mass of anchor in air (kg)

E_r modulus of elasticity in the longitudinal direction of the layer (N/m^2)

v_i impact velocity of anchor (m/s)

A cross-sectional striking area of anchor (m^2)

The calculated maximum penetration depth should not exceed 90% of the total thickness of the protection layer covering the tunnel using the 5% fractile value for E_r. The dynamic load factor (DLF) ratio of the static equivalent load on the tunnel roof to the triangular dynamic load pulse $F = m_w v_i / T_d$ may be obtained from figures using the minimum duration of impact $T_d = x / v_i$ (where x is calculated with the 95% fractal value for E_r), and the natural period T_0 of the affected element.

Ship Sinking

The primary sunken ship design case should be assumed to consist of a ship of the size approximating those using or expected to use the waterway. The imposed loading of a ship on the tunnel should be taken as an appropriate uniform loading over an area not exceeding the full width of the tunnel times a length as measured on the longitudinal axis of the tunnel of 100 feet (30 m). Collision impact loading should not be considered.

If appropriate, a secondary sunken ship design case should be assumed to consist of a smaller vessel, such as a ferry or barge, sinking and impacting the tunnel structure with the stem or sternpost in a manner similar to that of a dropped anchor. A static equivalent concentrated load of 225 kips (1,000 kN) working on an area of 3.3 x 6.6 ft^2 (1x2 m^2) directly on the tunnel roof should be considered.

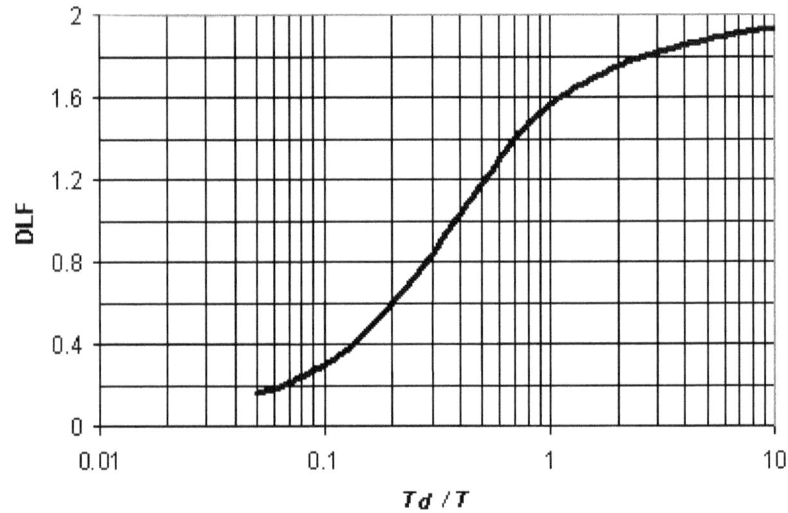

Figure: Graph of Dynamic Load Factor (DLF) Against Td/T

The intensity of uniformly distributed loading from a sunken ship should be determined by methods such as that outlined of the State-of-the-Art Report, 2nd Edition, International Tunnelling Association Immersed and Floating Tunnels Working Group, Pergamon, 1997. In the absence of data to the contrary, it may be assumed that the ship will exert a pressure of 1 ksf (50 kN/m^2).

Loads During Fabrication, Transportation and Placement

During fabrication, load effects caused by placement of concrete while the element is afloat or by settlements of the foundation (in case of concrete elements), and other items should be evaluated. Some of these loads may cause locked-in stresses that must be considered together with stresses due to external loads.

Particular care must be taken during the placement of concrete while an element is afloat to ensure not only that stresses stay within limits, but also that the deflected shape due to the weight of the new concrete is within acceptable limits. At all times when the element is afloat, stresses due to waves should be checked to ensure that all limit states are satisfied; the wave height and length used in design must be specified for each stage of construction and for towing so that measures can be taken to move the element to a place of safety when forecasts predict conditions that exceed allowable limits. If the freeboard is such that waves could run over the top of an element, this loading should also be taken into consideration. During transportation and while moored at the outfitting pier or elsewhere and even while in the fabrication yard, a tunnel element can be subject to wind loads that should be considered. The tunnel element may be suspended from lifting hooks during immersion and may be placed on temporary supports in the final location pending completion of the foundation. All limits states must be satisfied. Temporary supports if used should be released before backfill is placed. When adjacent tunnel elements are connected by shear keys, the effects due to relative differential settlements of each tunnel element during progressive backfilling operations must be taken into account.

The AASHTO LRFD specifications provides guidance for minimum load factors to be used when investigating loads that occur during construction. The following table reflects the load combinations and load factors to be used when evaluating immersed tunnel sections for construction loads.

Table 1: Construction Load Combinations

	DC	*EL*	*WS*	*CL*	*WA*
Strength I	1.25	1.00	0.00	1.5	1.00
Strength II	1.25	1.00	0.00	1.5	1.00
Strength III	1.25	1.00	1.25	1.5	1.00
Strength IV	1.25	1.00	0.00	1.5	1.00
Service I	1.00	1.00	1.25	1.5	1.03
Service IV	1.00	1.00	0.00	1.5	1.05

Structural Design

Historically there have been three basic methods used in the design of immersed tunnels:

- Service load or allowable stress design which treats each load on the structure equally in terms of its probability of occurrence at the stated value. The factor of safety for this method is built into the material's ability to withstand the loading.

- Load factor design accounts for the potential variability of loads by applying varying load factors to each load type. The resistance of the maximum capacity of the structural member is reduced by a strength reduction factor and the calculated resistance of the structural member must exceed the applied load.

Load and resistance factor design takes into account the statistical variation of both the strength of the structural member and of the magnitude of the applied loads. The fundamental LRFD equation can be found in paragraph of the AASHTO specification. This equation is:

$$\Sigma\eta_i\gamma_i Q_i \text{ d" } fR_n = R_r$$

In this equation, η is a load modifier relating to the ductility, redundancy and operation importance of the feature being designed. The load modifier η is comprised of three components:

η_D = a factor relating to ductility = 1.0 for immersed tunnels constructed with conventional details and designed in accordance with the AASHTO LRFD specification.

η_R = a factor relating to redundancy = 1.0 for immersed tunnel design. Typical cast in place and prestressed concrete structures are sufficiently redundant to use a value of 1.0 for this factor. Typical detailing using structural steel also provides a high level of redundancy.

η_I = a factor relating to the importance of the structure = 1.05 for immersed tunnel design. Tunnels usually are important major links in regional transportation systems. The loss of a tunnel will usually cause major disruption to the flow of traffic, hence the higher importance factor.

γ is a load factor applied to the force effects (Q) acting on the member being designed. Values for γ can be found in Table 11-1 above.

f is a resistance factor applied to the nominal resistance of the member (R) being designed. The resistance factors are given in the AASHTO LRFD specifications for each material in the section that covers the specific material. The following values to be used for f:

For Reinforced Concrete Box Structures:

f = 0.90 for flexure

f = 0.85 for shear

Since the walls, floors and roofs of immersed tunnel elements will experience axial loads, the resistance factor for compression must be defined. The value of f for compression can be found in the AASHTO LRFD specification given as:

f = 0.75 for compression

Values for f for precast construction are also given in Table 12.5.5-1. However, only rarely under unusual circumstances will a casting yard be set up to create the same controlled conditions that exist in a precast plant. Therefore, it is recommended that the f values given for cast-in-place concrete be used for the design of immersed tunnels.

Structural steel is also used in immersed tunnel construction. Structural steel is covered in the AASHTO LRFD specification. The following values for steel resistance factors:

For Structural Steel Members:

f_f = 1.00 for flexure

f_v = 1.00 for shear

f_c = 0.90 for axial compression for plain steel and composite members

R_r is the calculated factored resistance of the member or connection.

Structural Analysis

It is recommended that classical force and displacement methods be used in the structural analysis of concrete immersed tunnel elements. Other methods (as described below) may be used, but will rarely yield results that vary significantly from those obtained with the classical methods. The modelling should be based on elastic behaviour of the structure as per AASHTO. Steel immersed tunnels can also be analysed using the same structural model except that the efficiency of any curvature of the steel members will not be fully utilised. Most general purpose structural analysis programs have routines based on these principles for dimensional models.

Since all members of a concrete immersed tunnel element are subjected to bending and axial load, the secondary effects of deflections on the load affects to the structural members should be accounted for in the analysis. AASHTO LRFD specifications refer to this type of analysis as "large deflection theory". Most general purpose structural analysis software have provisions for including this behaviour in the analysis. If this behaviour is accounted for in the analysis, no further moment magnification is required. Alternatively finite element models can be used. These models can identify load sharing, account for secondary effects and identify load paths

Steel immersed tunnel elements, are complex assemblies of plates that might be curved, stiffeners and diaphragms. Simplifying these systems to the point where classical methods of analysis can be used often undermines the efficient use of materials that can result from complex load paths. Steel structures lend themselves well to sophisticated computer modelling such as finite element models. These models can identify load sharing, account for secondary effects and identify load paths. It is recommended that these models be used in the analysis of steel immersed tunnel sections.

The AASHTO LRFD specifications states that the mathematical model used to analyse the structure should include "...where appropriate, response characteristics of the foundation". The response foundation for an immersed tunnel element can be modelled through the use of a series of non-linear springs placed along the length of the bottom of the section. These springs are considered non-linear because they should be specified to act in only one direction, the downward vertical direction. This model will provide the proper distribution of loads to the bottom of the model and give the designer an indication if buoyancy is a problem. This indication is seen in observing the calculated displacements of the structure. A net upward displacement of the entire structure indicates that there is insufficient resistance to buoyancy.

Structural models for computer analysis are developed using the centroids of the structural members. Due to the thickness of the walls and the slabs of an immersed tunnel, it is important when calculating the applied loads, that the loads are calculated at the outside surface of the members. The load is then adjusted according to the actual length of the member as input.

Watertightness and Joints Between Elements

External Waterproofing of Tunnels

External waterproofing for tunnel elements should be considered for both steel tunnels and concrete tunnels. The waterproofing should envelop every part of the element exposed to soil or water with materials impervious to the surrounding waters. For steel tunnels the outer steel membrane would act as waterproofing membrane, while for concrete elements either steel or synthetic membrane should be used. For steel waterproofing membranes used on either concrete or steel elements, an appropriate corrosion protection and monitoring system should be used to ensure that the minimum design thickness is maintained during the life of the facility or an added sacrificial thickness should be provided. Non-structural steel membranes should be no less than 1/4 in (6 mm) thick.

The membrane should be watertight. Typical materials used for concrete elements include two coats of a spray-applied elasticized epoxy material; steel plates; and flexible PVC waterproofing sheet. Minimum thickness should be no less than 0.06 inch (1.5 mm), and anchored to the concrete using T-shaped ribs. The materials of the waterproofing system should have a proven resistance to the specific corrosive qualities of the surrounding waters and soils. The materials of the system should be flexible and strong enough to span any cracks that may develop during the life of the structure. Bituminous membranes are not recommended. The waterproofing system should preferably adhere at every point to the surfaces to which it is applied so that, if perforated at any one location, water may not travel under it to another. The areas of free water flow between the membrane and the underlying concrete in case of leakage should be limited to no more than 100sf (10m^2). For a steel tunnel, the membrane could be the external steel shell, provided that an adequate corrosion protection is provided either by cathodic protection or additional sacrificial thickness. Steel plates should be joined using continuous butt welds. All welds should be inspected and tested for soundness and tested for watertightness. Notwithstanding the provision of a membrane, the underlying structural concrete should be designed to be watertight.

Depending upon the type of waterproofing used, it may require protection on the sides and top of the tunnel elements to ensure that it remains undamaged during all operations up to final placement and during subsequent backfilling operations.

Joints

Joints between immersed tunnels elements can be classified as described below.

Immersion Joint (or Typical Joint) The immersion joint is the joint formed when a tunnel section is joined to a section that is already in place on the seabed. After placing the new element, and joining it with the previously placed element, the space between the bulkheads (dam plates) of the two adjoining elements is then dewatered. In order to dewater this space, a watertight seal must be made. A temporary gasket with a soft nose such as the Gina gasket is most often used.

In addition an omega seal is also provided after dewatering the joint from inside the joint.

Figure: Gina-Type Seal

For immersion joints, the primary compression or immersion seal is usually made of natural or neoprene rubber compounds. The most common cross-section used today is the "Gina" type. This consists of a main body with designed load/compression characteristics and an integral nose and seating ridge. The materials used should have a proven resistance to the specific corrosive qualities of the water and soils and an expected life no shorter than the design life of the tunnel unless the gasket is considered temporary. For flexible joints, a secondary seal is usually required in case of failure of the primary seal. It is usually manufactured from chloroprene rubber to an overall cross-section corresponding to that known as an "Omega" type, the materials having proven resistance against the specific corrosive qualities of the water and soils, oil, fungi and micro-organisms, oxygen, ozone and heat.

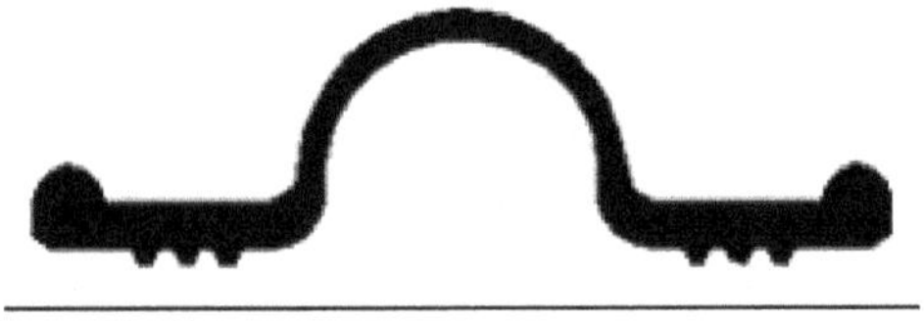

Figure: Omega Type Seal

A typical immersion joint. It is essential that immediately after dewatering of the chamber between the two bulkheads, an inspection of the primary seal is made so that any lack of watertightness can be remedied. Similarly, the secondary seal of a flexible joint should be pressure tested up to the expected maximum service pressure via a test pipe and valve to ensure that it too can function as required; after a successful testing, the chamber between the seals should be de-watered.

Closure or Final Joint: Where the last element has to be inserted between previously placed elements rather than appended to the end of the previous element, a marginal gap will exist at the secondary end. This short length of tunnel sometimes is completed as cast-in-place and is known as the closure or final joint.

The form of the closure or end joint is dependent on the sequence and method of construction. Closure joints may also be immersion joints, although details may need to be different. Potential options for the closure joints include:

Figure: Gina-type Immersion Gasket at Fort Point Channel, Boston, MA.

- Place the last element between two previously placed elements and dewater one joint between the newly placed element and the one of the previously placed elements. Then insert under water closure form plates and place tremie concrete around the closure joint to seal it. The joint can then be dewatered and interior concrete can be completed from within the joint. Other methods such as telescopic extension joints and wedge joints have been developed to make the closure joint similar to the immersion joint.
- Construct both end (terminal) joints first, lay the tunnel elements outwards from these and complete the immersed tunnel with a special closure (final) joint.
- Construct one terminal joint first and lay all the immersed tunnel elements outwards from that side and backfill over the top of the final element, using a soil-cement mixture (or other reasonably watertight material) in the vicinity of the second terminal joint. Construct the structures abutting the second terminal joint after the immersed tunnel is complete.

- Lay and complete the immersed tunnel with or without a special closure joint and backfill at the terminal elements using a soil-cement mixture (or other reasonably watertight material) in the vicinity of both terminal joints. Construct the structures abutting the both terminal joints after the immersed tunnel is complete.

Earthquake Joint This may be an immersion joint of special design to accommodate large differential movements in any direction due to a seismic event. It also applies to a semi-rigid or flexible joint strengthened to carry seismic loads and across which stressed or unstressed prestressing components may be installed.

Segment or Dilatation Joint Moveable segment joints must be able to transmit shear across the joint and well as allowing dilatation and rotation. The joints contain an injectable rubber-metal waterstop as well as neoprene and hydrophilic seals.

Design of Joints between Elements

All immersed tunnel joints must be watertight throughout the design life, and must accommodate expected movements caused by differences in temperature, creep, settlement, earthquake motions, method of construction, etc. Displacements in any direction should be limited so that the waterproof limits of a joint are not exceeded. Joint shear capability should take into account the influence of normal forces and bending moments on the shear capacity of the section; the design should take account of shear forces generated where the faces of the joints are not normal to the tunnel axis. Joints must be ductile in addition to accommodating longitudinal movements. Tension ties may be used to limit movement so that joints do not leak or break open, especially during a seismic event.

The axial compression of tunnel elements and bulkheads due to depth of immersion should be taken into account in determining joint dimensions at installation.

The design of primary flexible seals at tunnel joints must be designed to take into account the maximum deviations of the supporting frames relative to their theoretical location, the maximum deviation of the planes of the frames, and any relaxation of the seal. The seal is required to have a minimum compression of 3/8 inch (10 mm) greater than the compression required to maintain a seal. Just in case an initial seal is not obtained after immersion and joining, it may be advisable in some cases for the immersion joint to be designed so that a backup method of obtaining an initial seal is available. For flexible joints, a secondary seal (omega) capable of carrying the full water pressure should be fitted across the inside of the joint and should be capable of being inspected, maintained and replaced. The seal should be capable of absorbing the long-term movements of the joint. The secondary seals should be provided with a protective barrier against damage from within the tunnel. All joints in the tunnel should be finished to present a smooth surface.

The metal hardware in joints should have a design life adequate to fulfill its purpose throughout the design life of the joint. Nuts and bolts for primary and secondary seals should be stainless steel. Plate connections between elements should be corrosion-protected to ensure that the design life is obtained.

The mounting procedure or the mounting surface for the primary seal of immersion joints must allow for fine adjusting and trimming of the seal alignment in order to compensate for construction tolerances. It is recommended that the gasket be protected from accidental damage until the time of immersion. All embedded parts, fixings, including the bolts and their corrosion protection system, mating faces, clamping bars and other fixings, must have a design life at least equal to that of the tunnel structure. Where clamping bars and other fixings are used for the secondary seal, these need to have a design life at least equal to that of the secondary seal. The gasket assembly should have provision for injection in case of leakage.

Jacked Box Tunnelling

Jacked box tunnelling is a unique tunnelling method for constructing shallow rectangular road tunnels beneath critical facilities such as operating railways, major highways and airport runways without disruption of the services provided by those surface facilities or having to relocate them temporarily to accommodate open excavations for cut and cover construction. Originally developed from pipe jacking technology, jacked box tunnelling is generally used in soft ground at shallow depths and for relatively short lengths of tunnel, where TBM mining would not be economical or cut-and-cover methods would be too disruptive to overlying surface activities.

Jacked box tunnelling has mostly been used outside of United States (Taylor et al, 1998) until it was successfully applied to the construction of three short tunnels beneath a network of rail tracks at South Station in downtown Boston . These tunnels were completed and opened in 2003 as a part of the extension of Interstate I-90 for the Central Artery/Tunnel (CA/T) Project. The opening ceremony for the completed I-90 tunnels. Since CA/T Project represents the most significant application to date of the jacked box tunnelling in the US, it will be used to demonstrate the method throughout this Chapter.

Figure: Completed I-90 Tunnels

Basic Principles

The basic jacking sequence of jacked box tunnelling under an existing railway. The box structure is constructed on jacking base in a jacking pit located adjacent to one side of an existing railway. A tunnelling shield is provided at the front end of the box and hydraulic jacks are provided at the rear.

The box is advanced by excavating ground from within the shield and jacking the box forward into the opening created at the tunnel heading. In similar fashion to pipe jacking, lengths of tunnel that would exceed the capacities of jacks situated at the rear of the box structure can be successfully advanced into place by dividing the box structure into sections and establishing intermediate jacking stations. The box structure shown, is divided into two sections with an intermediate jacking station set up in between them.

In order to maintain support to the tunnel face, excavation and jacking normally carried out alternately in small increments, typically in the range of 2 to 4 feet. In most cases, the soft ground must be treated by means of ground improvement techniques such as ground freezing, jet grouting, etc.

Central Artery/Tunnel (CA/T) Project Jacked Box Tunnels

The use of the jacked box tunnelling method on the CA/T Project in Boston is described by van Dijk et al. (2000) and van Dijk et al., (2001). A major component of the CA/T project was the extension of Interstate I-90 eastward to Boston's Logan International Airport. This extension required three crossings of the network of tracks leading into South Station, a regional transportation hub used by Amtrak and the Massachusetts Bay Transportation Authority (MBTA) for hundreds of train movements daily. The critical surface use of the site, the large spans of the underground openings required to accommodate a multi-lane highway, the relatively shallow cover dictated by the roadway profile, and the poor soils in combination with the high groundwater level at the site led to tunnel jacking being selected as the preferred tunnelling method over staged cut-and-cover and conventional tunnelling techniques.

The three crossings of the tracks consisted of box structures for the eastbound lanes of I-90, the westbound lanes, and westbound exit ramp that provided access to Interstate I-93. The box structure for the I-90 EB lanes was the longest of the three, at 379 feet. It was constructed, with cross-sectional dimensions of 36 feet high by 79 feet wide, and a total weight of approximately 32,500 tons. The other two box structures were 38 feet high by 78 feet wide and were each constructed in two sections. The I-90 WB tunnel was 258 feet long and weighed approximately 27,000 tons, while the exit ramp tunnel was 167 feet long and weighed 17,000 tons.

Subsurface Condition and Ground Freezing: As shown, the geologic conditions through which the three box tunnel structures were jacked included (at the top of the subsurface profile) a layer of miscellaneous fill 20 to 25 feet thick, primarily a medium dense silty sand. This fill layer contained a number of obstructions related to the more than 150 years use of the site for rail road, industrial and waterfront infrastructure, which included granite block seawalls, rock filled timber cribwalls, brick and masonry structure foundations, a buried trackway, and an abandoned brick-lined sewer. Below the historic fill material was a deposit of weak organic sediments 10 to 15 feet thick, consisting of organic silt with some fine sand and peat. Underlying the organic layer were lenses of alluvial sand and inorganic silt deposits, generally less than 5 feet thick. The remaining part of the profile through which the tunnel boxes were jacked consisted of marine clay, consisting of clay and silt that was soft, except for the upper 15 feet, which was somewhat stronger and less compressible.

Groundwater at the site was generally 6 to 10 feet below track level, resulting in the tunnelling horizon in each case being completely submerged.

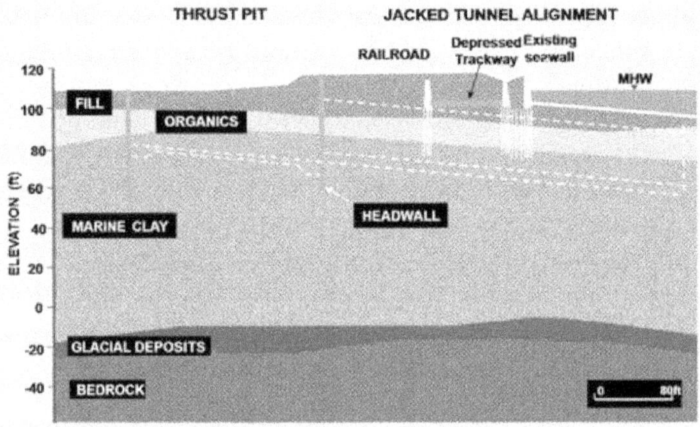

Figure: Generalised Subsurface Profile for the I-90 Jacked Box Tunnels

The success of the box jacking operation depended critically on maintaining the stability of the soils through which the tunnels passed. For the existing condition of weak soils below the groundwater table, shallow cover over the tunnel boxes, and the large spans required, there were serious concerns about loss of ground at the headings, the potential for significant settlement of the overlying track structures, and loss of alignment control during jacking. Therefore, ground improvement was required to enable the tunnelling to be performed effectively and safely. The original design for the tunnels had called for ground treatment consisting of a combination of dewatering and chemical grouting in the miscellaneous fill materials; horizontal jet grouting in the organic sediments; and soil nailing of the marine clay layer. The Contractor was concerned about the potential risks associated with the implementation of this combination of soil stabilisation methods, and consequently made a value engineering proposal to substitute ground freezing for all of the methods. This proposal was accepted, and a large scale freezing operation was performed that encompassed all three tunnel alignments.

Box Casting Operation: Each tunnel box structure was constructed in a "jacking pit" immediately adjacent to the west side of the South Station track network. The jacking pits were constructed by slurry wall methods, with post-tensioning of the sidewalls and the formation of a low-level improved soil "strut" formed by jet grouting below the base slab to reduce the number of bracing levels required so that the boxes could be constructed without interference from a cross-lot bracing system. The concrete base slab of each jacking pit was placed with a tight tolerance on grade, since it served as the casting surface for the box structures and also established the starting profile to ensure that the tunnel sections were jacked to the required alignment.

Figure: Tunnel Structure Construction Operation

After the base slab of each jacking pit was completed, a series of steel wire ropes were installed longitudinally on the slab and steel plates covering the entire footprint of the box sections were placed on the wire ropes. Shear studs were welded to the base plates to anchor the plates to the concrete slab, so that when jacking started, the frictional resistance that the jacks needed to overcome to move the box structures would result from steel plates sliding over steel wire ropes, rather than concrete sliding on concrete.

The structural design of these tunnel sections had to consider not only the long term loads from the overburden and railroad surcharge loads, but also the construction phase jacking loads. Each tunnel was constructed in sections to reduce the jacking forces required to move the tunnels into their final positions by using intermediate jacking stations in addition to the jacks positioned at the rear. To prevent soil from entering into the gap between adjacent box sections, a system of transversely continuous sliding overlapping steel "bridge" plates were used. Once jacking was completed, the jacks were removed and the intermediate jacking station areas were filled with concrete.

The external surfaces of the box structures could not be waterproofed because the waterproofing material would have been torn away during jacking. Water seepage control was achieved by using low permeability concrete mixes to construct the boxes and grouting the interface between the boxes and the surrounding ground through grout ports cast into the walls and roof slab after tunnelling and jacking were completed.

A cellular concrete shield was constructed at the front of each lead box section to support the excavation operation by establishing multiple access points to the face that could be closed off if stability problems developed. A beveled steel knife edge was provided at the perimetre of the shield that was flared a small amount to ensure that the opening into which the tunnel box structures would be jacked could be closely controlled, but also excavated large enough to prevent the boxes from getting stuck as they were pushed forward.

Tunnel Excavation: Mining of the frozen soils at the tunnel face, which had estimated uniaxial compressive strengths in the range of 700 to 1400 psi, was done primarily with roadheaders, working at two levels within the shield.

Figure: Excavation of the Frozen Ground at the Front of the Tunnel Shield

The roadheaders also proved to be effective at removing the numerous timber piles that were encountered. For removing masonry obstructions, which were firmly bound in place in the frozen soil mass, hydraulic hammers were used. The excavated material dropped to the bottom of the shield during the mining operation, where it was collected using a Gradall machine and a loader. A wheel-mounted scoop tram was used to shuttle the material to the rear of the tunnel box structure and dump it into a skip bucket, which was lifted out of the pit by crane and stockpiled for loading onto haul trucks.

Figure: Scoop Tram Loading Excavated Material into Skip Bucket for Removal

Based on typical mining production rates, stand-up time for the unsupported frozen ground, the volume of excavated material to be handled, the design of the jacking system, and the shift schedule, the Contractor determined that incremental excavation advance for efficient, consistent progression of the jacking operation was approximately 3 feet. Depending on the amount of obstructions encountered in a particular round, the advance rate achieved was generally one to two rounds per day, or 3 to 6 feet. At the completion of each excavation increment, the Contractor had to check the shield perimetre to ensure that all obstructions, including abandoned freeze pipes, were cut back sufficiently to be clear of the tunnel box.

Anti-Drag System As discussed previously, an anti-drag system was installed above and below the tunnel box structure to reduce the frictional resistance between the box structure and the surrounding ground. The system worked to even out the friction acting over the roof and bottom surface areas of the box, which contributed to alignment control during jacking, and also reduced the potential for surface settlement and lateral movement of the shallow overburden over the tunnel by separating the interface between the box concrete and the soil. This was achieved by installing a series of greased 3/4-inch diametre wire ropes that were anchored to the jacking pit and threaded through slots in the shield into the interior of the tunnel box structure, where they were stored on slings mounted on the soffit of the roof slab and on reels on the base slab located inside the tunnel. The system was configured so that as the tunnel moved forward, the wire ropes were run out from the storage units to cover the portion of the top and bottom surfaces of the box structure that was embedded in the ground beyond the thrust pit.

Tunnel Jacking Operation: At the completion of each excavation round, the tunnels structures were jacked into the space created at the face. This was accomplished by a group of 25 hydraulic jacks positioned at base slab level at the rear of the tunnel box, and additional groups of 26 to 32 jacks situated in the intermediate jacking stations. Each jack had a working capacity of 533 tons at a working pressure of 6100 psi, and could deliver a maximum thrust of 889 tons at a pressure of 10,200 psi. The maximum stroke of the rear jacks was 42 inches, while the stroke of the intermediate station jacks was limited to 16.5 inches. At each jacking station, the individual jacks were connected in nine clusters of 2 to 4 jacks each. This simplified the hydraulic control, and also enabled some horizontal steerage capability through variable operation of the clusters. The required thrust reaction for the jacks was transferred to a heavily reinforced concrete block wall at the rear of the jacking pit through a series of steel pipe sections referred to on the project as "packers." The loads exerted on the reaction block wall were in turn transferred into the surrounding ground through the pit base slab and rear wall.

Load and Structural Considerations

In most aspects, the structural loading and design considerations for jacked box tunnels are similar to those for cast-in-place cut and cover tunnels. However, in addition to the typical design, jacked box design can be dominated by two unique loads during construction: jacking thrust loads and interface drag loads.

Ground Drag Load and Anti-Drag System (ADS)

Ground drag, resulted from the contact pressures between soil and box structure is calculated and multiplied by appropriate friction factors, and is used to estimate drag loads at frictional interfaces; an appropriate adhesion value is used at the interface between the box and cohesive ground. Simplifying assumptions are made in developing ADS loads and modelling box/ADS/soil interaction, the validity of which is done by back-analyses of loads and other historical data. To reduce such an enormous drag load, an anti-drag system (ADS) is used to separate the external surface of the box from the adjacent ground during tunnel jacking.

The CA/T tunnels utilised an ADS consisting of an array of closely-spaced wire ropes which are initially stored within the box with one end of each rope anchored at the jacking pit. As the box advances, the ropes are progressively drawn out through guide holes in the shield and form a stationary separation layer between the moving box and the adjacent ground. The drag forces are absorbed by the ADS and transferred back to the jacking pit. In this manner the ground is isolated from drag forces and remains largely undisturbed.

Jacking Load

The ultimate bearing pressures on the face supports and on the shield perimetre are used to calculate the jacking load required to advance the shield. Jacking thrust is provided by means of specially built high capacity hydraulic jacking equipment. Jacks of 500 tons (4,448 kN) or more can be utilised on large tunnels. Jacks with a capacity of 533 tons at a working pressure of 6100 psi (42 MPa) were used in the I-90 tunnels. For jacking a large size road tunnel structure, multiple jacks are required to provide sufficient jacking thrust to counter the face pressure. In addition, using multiple jacks offers some steerage control redundant capacity in the event of possible underestimates of the required jacking loads.

Reaction to the jacking thrust developed is provided by either a jacking base or a thrust wall, depending on the site topography and the relative elevation of the tunnel. An example of a heavily reinforced thrust block wall is also shown. These temporary structures must in turn transmit the thrust into a stable mass of adjacent ground. A thrust wall is normally stabilised by passive ground pressure. In developing this reaction, the wall may move into the soil and this movement must be taken into account when designing the jacking system. When a thrust wall is used in a vertical sided jacking pit, care is required to ensure that movement of the thrust wall under load does not cause any lack of stability elsewhere in the pit.

Figure: Close Up of High Capacity Hydraulic Jacks, Reaction Blocks, and Packers

As the tunnel box and the jacks were gradually advanced away from the thrust wall, the Contractor needed to come up with a method to continue to transfer the jacking reaction force back into the thrust block wall. This was done by installing a series of 3-ft diametre structural steel pipe sections (i.e., packers) to bridge the gap between the jack pistons and the thrust block wall. Initial short packer sections installed once the tunnel box structure had been jacked away from the rear reaction block a distance exceeding the maximum stroke of the jacks. The packers were connected together with 1-inch thick diaphragm plates that were anchored to the base slab in the thrust pit. Three views of the packer installations are shown in Figures.

A jacking base is normally stabilised by shear interaction with the ground below and on each side. Where the interface is frictional, the interaction may be enhanced by surcharging the jacking base by means of pre-stressed ground anchors or compacted tunnel spoil. The jacking base is also stabilised by both the top and bottom ADS which are anchored to it.

Figure: Installation of Packer Sections and Connecting Diaphragm Plates

Ground Control

As discussed previously, the soft ground most likely will need to be pre-treated to provide sufficient stand-up time during jack tunnelling. In addition, ground may need to be stabilised in advance to control surface settlement must be controlled when tunnel jacking at such a shallow depth.

Ground Freezing for CA/T Project Jacked Tunnels

The Contractor made a value engineering proposal to replace the various soil stabilisation methods indicated in the Contract with ground freezing. This alternative approach offered several advantages, including the ability to completely stabilise the soil mass through which the tunnel box structures were jacked. In contrast, the horizontal jet grouting and soil nailing methods in the original design, would have required tunnel jacking to be interrupted periodically to permit installation of the ground improvement measures from the heading. Ground freezing also offered: 1), the advantages of improved face stability, which made breasting of shield compartments unnecessary, 2), better encapsulation of obstructions which otherwise had the potential to suddenly ravel into the heading when exposed, and 3) the avoidance of windows of untreated ground.

The freezing system was installed entirely from the ground surface overlying each tunnel alignment, within the track network. The Contactor selected a conventional brine freezing system, with an ammonia plant providing the refrigeration. In the freeze plant, ammonia gas was compressed, condensing it to a liquid, then evaporated to chill the brine to an average temperature range of -25°C to -30°C. The brine used to cool and eventually freeze the ground was circulated through circuits of vertical freeze pipes as shown schematically.

Each individual freeze pipe consisted of a 4.5-inch diametre steel pipe closed at the end, with a 2-inch diametre plastic pipe inserted in it that was open at the bottom. As shown, the chilled brine was pumped from a supply header line into the inner pipe, where it exited at the bottom and rose up in the annulus between the inner and outer pipe, cooling the surrounding ground in the process. At the top of the pipe, the brine was sent to the next freeze pipe for cooling circulation, as part of a circuit of 4 to 7 pipes. After passing through all of the pipes in the circuit, the brine was pumped back to the freeze plant for re-chilling through a return header pipe. The brine was circulated continuously in this manner through all of the circuits comprising the freeze zone in what was a closed system. The temperature of the ground mass was gradually lowered over a period of 4 to 5 months until the soil froze and an average target temperature of -10°C was reached.

The freeze pipes were installed within the track area using a sonic type drill rig, which used a vibratory coring bit to advance a starter hole through the miscellaneous fill material and any obstructions contained within it, and then vibrated the outer steel freeze pipe into place in a dry drilling process that displaced the underlying organic sediments and marine clay deposits. The drill rig was mounted on a turntable on the back of a high-rail truck vehicle, which provided flexibility for locating the pipes between rails and outside of the timber ties and switching and signal equipment.

Most of the drilling work was done at night by using a series of carefully coordinated track outages with the Railroad, and the sonic drilling method proved to be very effective for installing the freeze pipes quickly with relatively little drilling spoils being generated. Figure shows the system in operation while commuter trains continued to run through the freezing area. A total of nearly 1800 freeze pipes were used on the project to freeze over 3.5 million cubic feet of soil.

Figure: Ground Freezing System in Operation while Commuter Trains Run Through the Area

The ground freezing method was very effective at providing a stable face over the entire tunnel cross-sectional face area. The one significant disadvantage of the method was the expansion of water when it freezes caused the overlying track area to heave.

The amount of heave varied considerably over the alignment of each tunnel, depending on the variation in moisture content of the underlying soil profile. Typically the maximum deformation, which was monitored daily by detailed surveys of rail elevations, was in the range of 4 to 7 inches. The heave tapered to the original ground elevation over distances that extended laterally from tunnel centreline to approximate distances of about 50 to 70 feet beyond the edge of the tunnel box structure. The magnitude of this deformation required periodic re-profiling of the tracks by the Railroad to ensure that their rail geometry requirements for safe operation of their trains were maintained.

The temperature of the frozen soil mass was monitored by a series of temperature probes installed at each freeze site. After the target temperature was reached, the freeze system was adjusted to maintain that temperature, which controlled the stability of the soils at the tunnel face. As the excavation progressed for each tunnel, the freeze circuits were shutdown and the brine and inner pipes removed from the outer steel pipes, which were left in place. This progressive shut-down and dismantling of the freeze system was timed to avoid any significant warming of a section of the soil mass prior to it being exposed in the tunnel heading. When the abandoned steel freeze pipes were encountered, they were removed by cutting them out with a torch.

The Tunnel Designer should ensure that ground treatment measures do not in themselves cause an unacceptable degree of ground disturbance and surface movement.

Figure: Frozen Face Seen from Shield at Front of Jacked Box Structure

Face Loss

Design should also include provisions for controlling face loss which occurs when the ground ahead of the shield moves towards the tunnel as a result of reduction in lateral pressure in the ground at the tunnel face. With face loss, as the tunnel advances, a greater volume of ground is excavated than that represented by the theoretical volume displaced by the tunnel advance.

In cohesive ground, face loss is controlled by supporting the face at all times by means of a specifically-designed tunnelling shield and by careful control of both face excavation and box advance. The shield is normally divided into cells by internal walls and shelves which are pushed firmly into the face. Typically 0.5 ft (150mm) of soil is trimmed from the face following which the box is jacked forward 0.5 ft (150mm). This sequence is repeated until the tunnelling operation is complete, thus maintaining the necessary support to the face.

Over Cut

Design should also include provisions for controlling overcut in soft ground, by ensuring that the shield perimetre is kept buried and cuts the ground to the required profile. However, a degree of over-cut at the roof and sides beyond the nominal dimensions of the box is required for three reasons:

1. The hole through which the box travels must be large enough to accommodate irregularities in

the external surfaces of the box.

2. It is desirable to reduce contact pressures between the ground and the box, to reduce drag.
3. Overcutting may be required to fully remove obstructions at the perimetre of the shield.

The amount of over-cut required should be minimised if unnecessary ground disturbance and surface settlement is to be avoided. This demands that the external surfaces of the box be formed as accurately as possible. Typical forming tolerances are: ± 0.4 in (10mm) at the bottom and ± 0.6 in (15mm) at the walls and roof.

Bibliography

Abramson, L.W. and D.A. Bruce: *Ground Control and Improvement*, John Wiley and Sons, Inc., New York, 1994.

Barlow, Jewel B., Alan Pope, and William H. Rae, Jr.: *Low-Speed Wind Tunnel Testing*, John Wiley & Sons, Inc., New York, 1999.

Bickel, Kuesel and King: *Tunnel Engineering Handbook*; Chapman & Hall, N.Y., 1996.

Bieniawski, Z. T.: *"Engineering Rock Mass Classifications: A Complete Manual for Engineers and Geologists in Mining, Civil and Petroleum Engineering*, Wiley, NewYork, 1989.

__________: *Design Methodology in Rock Engineering*, Rotterdam, A.A., 1992.

__________: *Rock Mechanics Design in Mining and Tunneling*, A.A. Balkema, Rotterdam, 1984.

__________: *The Geomechanics Classification in Rock Engineering Applications, Proceedings*, ISRM, Montreux. A. A. Balkema, Rotterdam, 1979.

Brekke, T. L., and Ripley, B. D: *Design of Pressure Tunnels and Shafts*, J. A. Hudson, 1993.

Brierley, G.S. and Hatem, D.J.: *Design Build – Subsurface Projects*, Zeni House Books, NY, 2002.

Brierley, G.S.: *Subsurface Conditions Risk Management for Design and Construction Management Professionals*, John Wiley & Sons, Inc., NY, 1998.

Brown, E. T.: *Rock Characterization Testing and Monitoring: ISRM Suggested Methods*, Pergammon Press, NJ, 1981.

Chen and Lui: *Handbook of Structural Engineering*, CRC Press, NY, 2005.

Collins, Luke, James Flint, Ben Morgan, Clint Witchalls, and Chris Woodford: *Cool Stuff and How It Works*. DK Publishing, Inc., New York, 2005.

Cording, E.J., and Deere, D.U.: *Rock Tunnel Supports and Field Measurement*, AIME, Chicago, 1972.

Critchfield, J. W.: *Tunneling in Gassy Ground*, Proceedings, RETC, AIME, New York, 1985.

Cundall, P. A., and Hart, R. D.: *Numerical Modeling of Discontinuities,* Pergammon Press, ML, 1993.

Cundall, P. A.: *Explicit Finite Difference Methods in Geomechanics, Numerical Methods in Engineering,* Blacksburg, Virginia, 1976.

Daugherty, C.W.: *Monitoring of Movements Above Large Shallow Jacked Tunnels,* ASCE Jacked Tunnel Design & Construction, Boston, 1998.

Dowding, C. H.: *Blast Vibration Monitoring and Control,* Prentice Hall, New York, 1985.

Dunnicliff, J.: *Geotechnical Instrumentation for Monitoring Field Performance,* John Wiley & Sons, New York, 1998.

Einstein, H.H.: *Design and Analysis of Underground Structures in Swelling and Squeezing Rocks,* R.S. Sinha, New York, 1989.

Gall, V., Munfah, N. and Cerulli, D Zeidler, K.: *Shotcrete for Tunnel Final Linings – Design and Construction Considerations,* North American Tunneling, NAT, 2004.

Goodman, R. E., and Shi, G. H.: *Block Theory and its Application to Rock Engineering,* Prentice-Hall, New Jersey, 1985.

Hansmire, W. and Monsees, J.E.: *Evolving Practices for Tunnel Contracts in the United States,* Elsevier, London, 2006,

Hemphill, G. B.: *Mechanical Excavation System: Selection, Design, and Performance in Weak Reek*; University of Idaho, Moscow, 1990.

Heuer, R.E.: *Estimating Rock Tunnel Water Inflow-II," Proceedings of RECT, 2005,* Seattle, SME, 2005.

Hoek, E., and Brown, E. T.: *Underground Excavations in Rock,* Institution of Mining and Metallurgy, London, England, 1980.

Hoek, E., Kaiser, P.K. and Bawden. W.F.: *Support of Underground Excavations in Hard Rock,* Rotterdam, Balkema, 1995.

Keifer, O.: *Multiple Layer Shotcrete Lining,* ACI Publications, NY, 1966.

Kuesel, T.: *Immersed Tube Tunnels: Concept Design and Construction,* Boston Society of Civil Engineers, Spring, 1986.

Kuhlmann, D.: *Wayss & Freitag am Donnerstag, Gespraeche mit Wissenschaft und Praxis,* Frankfurt, a.M., Germany, 1086.

Leonards, G.: *Foundation Engineering,* McGraw-Hill, New York, 1962.

Lowe, J. III and Zaccheo, P.F.: *Subsurface Explorations and Sampling, Foundation Engineering Handbook,* Van Nostrand Reinhold, NY, 1991.

Monsees, J. E., and J. L. Merritt : *Lifeline Earthquake Engineering,* ASCE, New York, 1991.

Monsees, J. E.: Soft Ground Tunneling, *Tunnel Engineering Handbook, Bickel,* Chapman and Hall, New York City, 1996.

Paretsky, Sara: *Tunnel Vision,* Dell Publishing, New York, 1994.

Pariseau, W. G.: *Design Analysis in Rock Mechanics,* CRC Press, London, 2006.

Peck, R. B.: *Foreword to Geotechnical instrumentation for Monitoring Field Performance*, Wiley, New York, 1988.

__________: *Influence of Non-technical Factors on the Quality of Embankment Dams, Embankment Dam Engineering,* Wiley, New York, 1973.

__________: *Judgment in Geotechnical Engineering: The Professional Legacy of Ralph B. Peck, J. Dunnicliff and D. U. Deere,* Wiley, New York, 1972.

Peterson, DL, C.R. Nelson, B.K. Nelson, and B.D. Wagener: *Design and construction of the Lafayette Bluff Tunnel*, Proceedings RETC, Seattle, 1991.

Phien-Wej, N., and EJ. Cording: *Sheared Shale Response to Deep TBM Excavation*, Elsevier, New York, 1991.

Powers, J. P.: *Construction Dewatering-A Guide to Theory and Practice*, John Wiley and Sons, New York, 1981.

__________: *Construction Dewatering-New Methods and Applications,* John Wiley and Sons, New York, 1992.

__________: *Groundwater Control in Tunnel Construction,* ASCE-AIME, New York, 1972.

Procter, R. V., and White, T. L.: Rock Tunneling with Steel Supports, The Commercial Shearing and Stamping Company, Ohio, 1946.

Proctor, R.Y. and T. L. White: *Earth Tunneling with Steel Supports*, Commercial Shearing, Inc., Youngstown, OH, 1977.

Ricketts, Loftin, and Merritt: *Standard Handbook for Civil Engineers*, McGraw-Hill, NY, 2003.

Roark, R.J., and Yomg,W.C.: *Formulas for Stvess and Strain*, McGraw-Hill, New York, 1975.

Russell, H. A.: *The Inspection and Rehabilitation of Transit Tunnels,* Parsons Brinckerhoff Quade & Douglas Inc. New York, 1992.

Shuster, J.A.: *Ground Freezing for Soft Ground Shaft Sinking*, Prentice-Hall, New York, 1985.

Skempton, A.W ., and D. H. MacDonald: *The Allowable Settlements of Buildings, Proceedings*, Institute of Civil Engineers, London, 1956.

Smith, H. C.: *The Illustrated Guide to Aerodynamics*, TAB Books, New York, 1992.

Stack, B.: *Handbook of Mining and Tunneling Machinery*, Wiley, New York, 1982.

Stillborg, B.: *Professional Users Handbook for Rock Bolting*, TransTech Publications, Atlas Copco., 1986.

Sullivan, Walter: *Progress in Technology Revives Interest In Great Tunnels*, New York Times, 1986.

Teng, W.C.: *Foundation Design*, Prentice-Hall, Inc., Englewood, NJ, 1962.

Terzaghi, K, and R.B. Peck: *Soil Mechanics in Engineering Practice*, John Wiley and Sons, New York, 1967.

Terzaghi, K.: *Earth Tunneling with Steel Supports*, Commercial Shearing and Stamping Co., Youngstown, OH, 1977.

__________: *Geologic Aspects of Soft Ground Tunneling,* John Wiley & Sons., New York, 1950.

Timoshenko, S.: Theory of Elastic Stability, McGraw-Hill Book Co., New York, 1936.

Tschebotarioff, G.P.: *Retaining Structures, Foundation Engineering*, McGraw-Hill, New York, 1962.

Van Dijk, P.A., Taylor, S., and Rice P.M.: *Box Jacking in Boston*, North American Tunneling, Boston, 2000.

Vandewalle, M., 2005: "Tunneling is an Art". NV Bekaert SA.

Wittke, W.: *Felsmechanik,* Springer, Berlin, 1984.

Wood, A.M.: *Tunneling, Management,* E & FN Spon, London, 2000.

Xanthakos, P.P.: *Slurry Walls*, McGraw-Hill, New York, 1979.

Zosen, H.: *Hitachi Zosen's Shield Tunneling Machines*, Company Brochure, Tokyo, 1984.

Index

F

G

H

J

L

M

O

P

R

S

T

❑❑❑

www.ingramcontent.com/pod-product-compliance
Ingram Content Group UK Ltd.
Pitfield, Milton Keynes, MK11 3LW, UK
UKHW021328190726
13845UKWH00004B/79